STUDENT WORKBOOK

for

Argument-Driven Inquiry

in

Fourth-Grade Science

Three-Dimensional Investigations

STUDENT WORKBOOK

for
Argument-Driven Inquiry
in
Fourth-Grade Science
Three-Dimensional Investigations

Victor Sampson and Ashley Murphy

National Science Teachers Association
Arlington, Virginia

National Science Teachers Association

Claire Reinburg, Director
Rachel Ledbetter, Managing Editor
Andrea Silen, Associate Editor
Jennifer Thompson, Associate Editor
Donna Yudkin, Book Acquisitions Manager

ART AND DESIGN
Will Thomas Jr., Director

PRINTING AND PRODUCTION
Catherine Lorrain, Director

NATIONAL SCIENCE TEACHERS ASSOCIATION
David L. Evans, Executive Director

1840 Wilson Blvd., Arlington, VA 22201
www.nsta.org/store
For customer service inquiries, please call 800-277-5300.

Copyright © 2019 by Argument-Driven Inquiry, LLC.
All rights reserved. Printed in the United States of America.
22 21 20 19   4 3 2 1

NSTA is committed to publishing material that promotes the best in inquiry-based science education. However, conditions of actual use may vary, and the safety procedures and practices described in this book are intended to serve only as a guide. Additional precautionary measures may be required. NSTA and the authors do not warrant or represent that the procedures and practices in this book meet any safety code or standard of federal, state, or local regulations. NSTA and the authors disclaim any liability for personal injury or damage to property arising out of or relating to the use of this book, including any of the recommendations, instructions, or materials contained therein.

PERMISSIONS
Book purchasers may photocopy, print, or e-mail up to five copies of an NSTA book chapter for personal use only; this does not include display or promotional use. Elementary, middle, and high school teachers may reproduce forms, sample documents, and single NSTA book chapters needed for classroom use only. E-book buyers may download files to multiple personal devices but are prohibited from posting the files to third-party servers or websites, or from passing files to non-buyers. For additional permission to photocopy or use material electronically from this NSTA Press book, please contact the Copyright Clearance Center (CCC) (*www.copyright.com*; 978-750-8400). Please access *www.nsta.org/permissions* for further information about NSTA's rights and permissions policies.

Cataloging-in-Publication Data are available from the Library of Congress.
LCCN: 2018059472
ISBN: 978-1-68140-570-4
e-ISBN: 978-1-68140-571-1

Contents

SECTION 1 - Introduction and Investigation Safety

Introduction 3

Safety Rules 5

Safety Acknowledgment Form 7

SECTION 2 - Energy

Investigation 1. Energy of Motion: How Does Changing the Speed of a Marble Affect the Energy of That Marble?

    Investigation Log 11

    Investigation Report Grading Rubric 24

    Checkout Questions 25

Investigation 2. Energy Transfer Through Collisions: How Does the Energy of a Moving Ball Change After It Collides With Objects That Have Different Masses?

    Investigation Log 27

    Investigation Report Grading Rubric 40

    Checkout Questions 41

Investigation 3. Energy Transfer Between Liquids: How Can We Predict How Much the Temperature of 100 ml of Water Will Change When We Add Different Amounts of Hot Water to It?

    Investigation Log 43

    Investigation Report Grading Rubric 56

    Checkout Questions 57

Investigation 4. Energy Transfer by Light: How Does the Distance Between a Light Source and an Object Affect the Amount of Energy an Object Receives From the Light Source?

    Investigation Log 59

    Investigation Report Grading Rubric 72

    Checkout Questions 73

Investigation 5. Electric Currents: How Does Adding More Batteries or Bulbs to a Closed Circuit Affect the Brightness of a Single Bulb Within That Circuit?

    Investigation Log 75

    Investigation Report Grading Rubric 88

    Checkout Questions 89

Contents

Investigation 6. Energy Transfer by Electric Currents: How Can We Make an Electric Car Move Faster?
- Investigation Log 91
- Investigation Report Grading Rubric 104
- Checkout Questions 105

SECTION 3 - Waves and Their Application in Technologies for Information Transfer

Investigation 7. Production of Waves: How Does Changing the Mass of an Object That Is Dropped Into Water Affect the Height of the Resulting Wave and the Position of a Floating Toy Boat?
- Investigation Log 109
- Investigation Report Grading Rubric 122
- Checkout Questions 123

Investigation 8. Characteristics of Sound Waves: How Can We Use the Properties of Waves to Explain Different Sounds?
- Investigation Log 125
- Investigation Report Grading Rubric 139
- Checkout Questions 140

Investigation 9. Light and Reflection: What Is the Relationship Between the Location of an Object and the Location of Its Image in a Mirror?
- Investigation Log 143
- Investigation Report Grading Rubric 156
- Checkout Questions 157

Investigation 10. Patterns and the Transfer of Information: How Can We Use an Electric Circuit to Accurately Transfer Information About the Content of a Picture?
- Investigation Log 159
- Investigation Report Grading Rubric 173
- Checkout Questions 174

Contents

SECTION 4 - From Molecules to Organisms: Structures and Processes

Investigation 11. Structures of Plants: How Does Water Move From the Roots to the Leaves of a Plant?

    Investigation Log 177

    Investigation Report Grading Rubric 190

    Checkout Questions 191

Investigation 12. Structures of Animals: How Should We Classify the Unknown Organisms?

    Investigation Log 193

    Investigation Report Grading Rubric 206

    Checkout Questions 207

Investigation 13. Information From Senses: What Type of Environmental Conditions Do Mealworms Seek Out?

    Investigation Log 209

    Investigation Report Grading Rubric 222

    Checkout Questions 223

SECTION 5 - Earth's Place in the Universe and Systems

Investigation 14. Movement of Water: Why Can We See the Roots of Trees That Grow Near Rivers or Streams?

    Investigation Log 227

    Investigation Report Grading Rubric 240

    Checkout Questions 241

Investigation 15. Earth's Features: Why Do Large Waves Often Block the Entrance to Some Harbors in New Zealand?

    Investigation Log 243

    Investigation Report Grading Rubric 256

    Checkout Questions 257

    Image Credits 259

Section 1
Introduction and Investigation Safety

Introduction

Science is much more than a collection of facts or terms. Science is a way to figure out how the world works and why it works that way. In science, we use core ideas, crosscutting concepts, and practices to figure things out.

The core ideas of science include the theories and laws that scientists in different fields of science use to develop new explanations for why or how things happen. For example, life scientists use the core idea of heredity to figure out how a trait is passed down from parent to child, and earth scientists use the core idea of weather and climate to figure out why it rains a lot in some places or why it is cooler during different parts of the year.

The crosscutting concepts of science are themes that have value in every field of science as a way to help scientists understand why or how things happen. For example, life scientists and earth scientists both look for patterns and cause-and-effect relationships when they are trying to develop explanations for why or how things happen. These crosscutting concepts help them think about what it is important to think about or look for during an investigation.

Finally, scientists use the practices of science to develop and refine new ideas. The practices include such things as asking and answering questions, planning and carrying out investigations, analyzing and interpreting data, and obtaining, evaluating, and communicating information. One of the most important practices of science is arguing from evidence. These core ideas, crosscutting concepts, and practices of science are important in science because most, if not all, scientists use them to figure out how the world works and why it works that way.

These core ideas, crosscutting concepts, and scientific practices are important for you to learn while you are in school. When you understand these core ideas, crosscutting concepts, and practices, it is easier to make sense of what scientists try to do. It is also easier to talk about what is currently going on in science with other people and to evaluate what you read or hear about science in the news. Once you learn the core ideas, crosscutting concepts, and practices of science, you will also have the knowledge and skills that you need to continue to learn about science outside school or to enter a career in science, technology, or engineering.

The investigations that are included in this book are designed to help you learn the core ideas, crosscutting concepts, and practices of science. During each investigation, you will have an opportunity to use a core idea, a crosscutting concept, and several practices of science to figure something out. Your teacher will introduce each investigation by giving you a task to accomplish and a guiding question to answer. You will then work as part of a group to plan and carry out an investigation to collect the data that you need to answer that question. From there, your group will develop an argument that includes an answer to the guiding question. You will then

Introduction

have an opportunity to share your argument with your classmates and critique their arguments, much like professional scientists do.

Next, you will be asked to revise your draft argument based on their feedback. You will then be asked to write an investigation report on your own to share what you learned. Your classmates will review this report before you submit it to your teacher for a grade.

As you complete more and more investigations in this book, you will not only learn the core ideas associated with each investigation but will also get better at using the crosscutting concepts and practices of science to understand the natural world.

Safety Rules

You will be doing many different investigations this year. You will use different materials, tools, and chemicals during these investigations. It is very important that you follow these 11 safety rules to keep you and your classmates from getting hurt when you use them:

1. Act in a responsible manner at all times.
2. Do not eat, drink, or chew gum.
3. Do not touch, taste, or smell any materials, tools, or chemicals without permission.
4. Wear safety goggles at all times.
5. Take care of the materials and tools that you use.
6. Tell your teacher about any accidents as soon as they happen.
7. Wear appropriate clothing, including closed-toed shoes and pants. Clothes should not be loose, baggy, or bulky. Use hair ties to keep long hair out of the way.
8. Keep work areas clean and neat at all times.
9. Clean work area and any materials or tools at the end of the investigations.
10. Wash your hands with soap and water at the end of the investigations.
11. Follow the teacher's directions at all times.

Your teacher may provide some more rules to follow during some of your investigations, but you must always follow these 11 rules. Your teacher will also warn you about anything that you need to keep in mind to stay safe as you work though an investigation.

Your teacher will go over a safety acknowledgment form with you before you start your first investigation. You will need to sign this safety acknowledgment form so your teacher knows that you understand all the safety rules and agree to follow them. Be sure to show this form to a parent or guardian after you have signed it. Your parent or guardian will also need to read and sign the safety acknowledgment form before your teacher will allow you to participate in any investigations.

Safety Acknowledgment Form

I know that it is very important to be as safe as I can during an investigation. My teacher has told me how to be safer in science. I agree to follow these 11 safety rules when I am working with my classmates to figure things out in science:

1. I will act in a responsible manner at all times. I will not run around the classroom, throw things, play jokes on my classmates, or be careless.
2. I will never eat, drink, or chew gum.
3. I will never touch, taste, or smell any materials, tools, or chemicals without permission.
4. I will wear my safety goggles at all times during the activity setup, hands-on work, and cleanup.
5. I will do my best to take care of the materials and tools that my teacher allows me to use.
6. I will always tell my teacher about any accidents as soon as they happen.
7. I will always dress in a way that will help keep me safer. I will wear closed-toed shoes and pants. My clothes will not be loose, baggy, or bulky. I will also use hair ties to keep my hair out of the way while I am working if my hair is long.
8. I will keep my work area clean and neat at all times. I will put my backpack, books, and other personal items where my teacher tells me to put them and I will not get them out unless my teacher tells me that it is okay.
9. I will clean my work area and the materials or tools that I use.
10. I will wash my hands with soap and water at the end of the activity.
11. I will follow my teacher's directions at all times.

_____   _____   _____
Print Name   Signature   Date

I have read and reviewed the 11 investigation safety rules with my child. He or she understands how important it is to follow safety rules in science and has agreed to follow these safety rules at all times. I give my permission for my child to participate in the investigations this year.

_____   _____   _____
Parent or Guardian Name   Parent or Guardian Signature   Date

Section 2
Energy

Investigation Log

Investigation 1

Energy of Motion: How Does Changing the Speed of a Marble Affect the Energy of That Marble?

Introduction

The concept of energy is important for all of us to understand because energy is what makes things change or move. For example, it takes energy to run the engine of a car or to cook food. It also takes energy to lift heavy rocks or to knock over a stack of blocks. Take a moment to explore what happens when you use a rubber band to launch a marble at a plastic cup. Be sure to pull the rubber band back by different amounts and pay attention to how much the cup moves each time. As you explore what happens, keep track of what you observe and what you are wondering about in the boxes below.

Things I OBSERVED ...	Things I WONDER about ...

Student Workbook for Argument-Driven Inquiry in **Fourth-Grade Science:** Three-Dimensional Investigations

Investigation Log

The motion of an object will change when energy transfers into or out of it. When you pulled on the rubber band, you transferred energy to it and made it move. When you let go of the rubber band, the energy you added to the rubber band transferred to the marble. The transfer of energy from the rubber band to the marble made the marble start to move across the table. When the moving marble hit the cup that was not moving, energy from the marble transferred to the cup. The transfer of energy from the marble to the cup made the cup start to move. There is always energy present whenever there is a moving object. In this investigation, you will need to figure out how changing the speed of a moving marble (a *cause*) affects the energy that the marble has because of its motion (the *effect*).

You will be able to examine the relationship between speed and energy and how energy can transfer from one object to another by measuring how far a cup moves after a marble hits it. You can change the speed of a marble by placing it in different spots on a ramp. You can measure the speed of a marble by timing how long it takes the marble to travel a specific distance. The distance the cup moves after it is hit by the marble will represent the amount of energy in the marble. You can determine how much energy a marble has by measuring how far a cup moves after the marble hits it because energy transfers between objects during a collision. The more energy that is transferred from the marble to the cup during the collision, the more the cup will move after the collision.

Things we KNOW from what we read …	What we will NEED to figure out …

National Science Teachers Association

Investigation 1. Energy of Motion:
How Does Changing the Speed of a Marble Affect the Energy of That Marble?

Your Task

Use what you know about the transfer of energy and cause-and-effect relationships to design and carry out an investigation to examine how a change in the speed of an object affects the energy of the object.

The *guiding question* of this investigation is, **How does changing the speed of a marble affect the energy of that marble?**

Materials

You may use any of the following materials during your investigation:

- Safety glasses or goggles (required)
- Marble
- Plastic cup (3 ounces)
- Grooved ruler
- Meterstick
- 2 plastic toy racecar tracks
- Track connector
- 4" × 2" x 2" wood block
- Stopwatch

Safety Rules

Follow all normal safety rules. In addition, be sure to follow these rules:

- Wear sanitized safety glasses or goggles during setup, investigation activity, and cleanup.
- Do not throw objects or put any objects in your mouth.
- Keep fingers and toes away from moving objects.
- Wash your hands with soap and water when you are done collecting the data.

Plan Your Investigation

Prepare a plan for your investigation by filling out the chart that follows; this plan is called an *investigation proposal*. Before you start developing your plan, be sure to discuss the following questions with the other members of your group:

- What information do we need to find a relationship between a **cause** and an **effect**?
- How can we **track the transfer of energy between objects**?

Investigation Log

Our guiding question:

This is a picture of how we will set up the equipment:

We will collect the following data:

These are the steps we will follow to collect data:

I approve of this investigation proposal.

_____   _____
Teacher's signature   Date

Investigation 1. Energy of Motion:
How Does Changing the Speed of a Marble Affect the Energy of That Marble?

Collect Your Data

Keep a record of what you measure or observe during your investigation in the space below.

Analyze Your Data

You will need to analyze the data you collected before you can develop an answer to the guiding question. To do this, create a graph that shows the relationship between the cause and the effect.

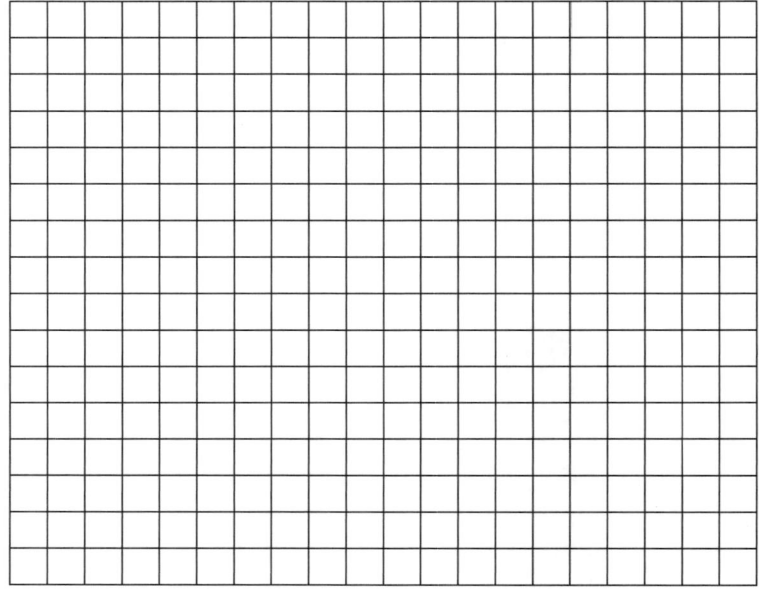

Student Workbook for Argument-Driven Inquiry in **Fourth-Grade Science:** Three-Dimensional Investigations

Draft Argument

Develop an argument on a whiteboard. It should include the following:

1. A *claim:* Your answer to the guiding question.
2. *Evidence:* An analysis of the data and an explanation of what the analysis means.
3. A *justification of the evidence:* Why your group thinks the evidence is important.

The Guiding Question:	
Our Claim:	
Our Evidence:	Our Justification of the Evidence:

Argumentation Session

Share your argument with your classmates. Be sure to ask them how to make your draft argument better. Keep track of their suggestions in the space below.

Ways to IMPROVE our argument …

Investigation 1. Energy of Motion:
How Does Changing the Speed of a Marble Affect the Energy of That Marble?

 **Draft Report**

Prepare an *investigation report* to share what you have learned. Use the information in this handout and your group's final argument to write a *draft* of your investigation report.

Introduction

We have been studying _____ in class.

Before we started this investigation, we explored _____

_____

_____

We noticed _____

_____

_____

My goal for this investigation was to figure out _____

_____

_____

The guiding question was _____

_____

_____

Method

To gather the data I needed to answer this question, I _____

_____

_____

_____

_____

Investigation Log

I then analyzed the data I collected by _____

_____

_____

_____

Argument

My claim is _____

_____

_____

_____

The graph below shows _____

_____

_____

_____

_____

_____

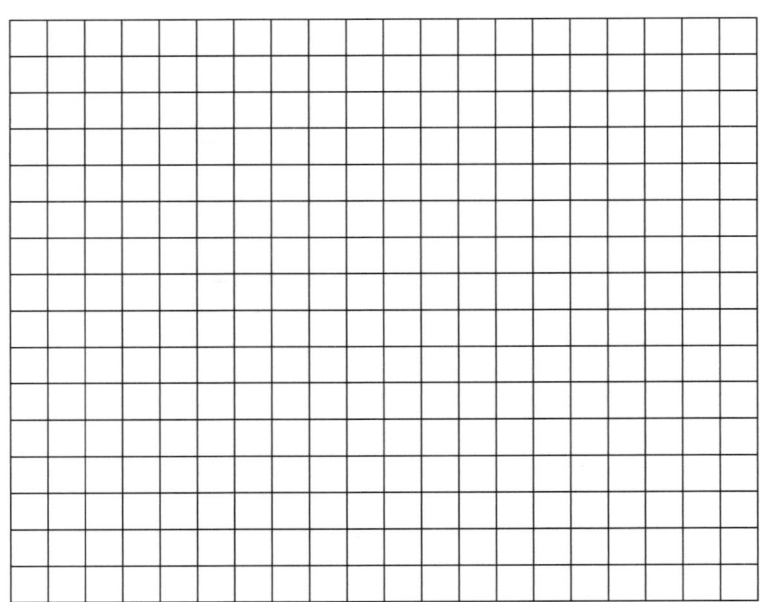

Investigation 1. Energy of Motion:
How Does Changing the Speed of a Marble Affect the Energy of That Marble?

This analysis of the data I collected suggests _____

_____

_____

_____

_____

This evidence is based on several important scientific concepts. The first one is _____

_____

_____

_____

_____

_____

_____

 **Review**

Your friends need your help! Review the draft of their investigation reports and give them ideas about how to improve. Use the *peer-review guide* that begins on the next page to guide your review.

Peer-Review Guide

Section 1: The Investigation	Reviewer Rating		
1. Did the author do a good job of explaining what the investigation was about?	☐ No	☐ Almost	☐ Yes
2. Did the author do a good job of making the **guiding question** clear?	☐ No	☐ Almost	☐ Yes
3. Did the author do a good job of describing what he or she did to **collect data?**	☐ No	☐ Almost	☐ Yes
4. Did the author do a good job describing **how** he or she **analyzed** the data?	☐ No	☐ Almost	☐ Yes
Reviewers: If your group gave the author any "No" or "Almost" ratings, please give the author some advice about what to do to improve this part of his or her investigation report.			

Section 2: The Argument	Reviewer Rating		
1. Does the author's claim provide a clear and detailed **answer** to the guiding question?	☐ No	☐ Almost	☐ Yes
2. Did the author support his or her claim with **scientific evidence?** Scientific evidence includes analyzed data and an explanation of the analysis.	☐ No	☐ Almost	☐ Yes
3. Does the **evidence** that the author uses in his or her argument **support the claim?**	☐ No	☐ Almost	☐ Yes
4. Did the author include enough **evidence** in his or her argument?	☐ No	☐ Almost	☐ Yes
5. Did the author do a good job of **explaining why the evidence** is important (why it matters)?	☐ No	☐ Almost	☐ Yes
6. Is the content of the argument **correct** based on the science concepts we talked about in class?	☐ No	☐ Almost	☐ Yes
Reviewers: If your group gave the author any "No" or "Almost" ratings, please give the author some advice about what to do to improve this part of his or her investigation report.			

Continued

Investigation 1. Energy of Motion:
How Does Changing the Speed of a Marble Affect the Energy of That Marble?

Section 3: Mechanics	Reviewer Rating		
1. *Grammar:* Are the sentences complete? Is there proper subject-verb agreement in each sentence? Are there no run-on sentences?	☐ No	☐ Almost	☐ Yes
2. *Conventions:* Did the author use proper spelling, punctuation, and capitalization?	☐ No	☐ Almost	☐ Yes
3. *Word Choice:* Did the author use the right words in each sentence (for example, *there* vs. *their*, *to* vs. *too*, *then* vs. *than*)?	☐ No	☐ Almost	☐ Yes

Reviewers: If your group gave the author any "No" or "Almost" ratings, please give the author some advice about what to do to improve the writing mechanics of his or her investigation report.

General Reviewer Comments

We liked …

We wonder …

Investigation Log

Write Your Final Report

Once you have received feedback from your friends about your draft report, create your final investigation report in the space that follows.

Introduction

_____
_____
_____
_____
_____
_____
_____
_____

Method

_____
_____
_____
_____
_____
_____
_____
_____
_____
_____
_____
_____

Investigation 1. Energy of Motion:
How Does Changing the Speed of a Marble Affect the Energy of That Marble?

Argument

Investigation Log

Investigation Report Grading Rubric

Section 1: The Investigation	Score		
	Missing	Somewhat	Yes
1. The author explained what the investigation was about.	0	1	2
2. The author made the **guiding question** clear.	0	1	2
3. The author **described** what he or she did to **collect data.**	0	1	2
4. The author described **how** he or she **analyzed** the data.	0	1	2

Section 2: The Argument	Score		
	Missing	Somewhat	Yes
1. The claim includes a clear and detailed **answer** to the guiding question.	0	1	2
2. The author used **scientific evidence** to support the claim. Scientific evidence includes analyzed data and an explanation of the analysis.	0	1	2
3. The evidence **supports the claim.**	0	1	2
4. The author included enough **evidence** in his or her argument.	0	1	2
5. The author **explained why the evidence** is important.	0	1	2
6. The content of the argument is **correct.**	0	1	2

Section 3: Mechanics	Score		
	Missing	Somewhat	Yes
1. *Grammar:* The sentences are complete. There is proper subject-verb agreement in each sentence. There are no run-on sentences.	0	1	2
2. *Conventions:* The author used proper spelling, punctuation, and capitalization.	0	1	2
3. *Word Choice:* The author used the right words in each sentence (e.g., *there* vs. *their, to* vs. *too, then* vs. *than*).	0	1	2

Teacher Comments

Here are some things I really liked about your report ...	Here are some things I think you could do next time to make your report even better ...

Total: _____ /26

Checkout Questions

Investigation 1. Energy of Motion

Use the following information to answer questions 1–3.

The picture below shows a ball that is rolling on a table toward a cup.

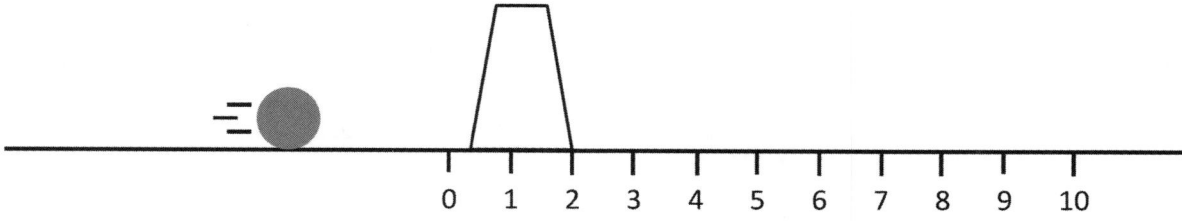

The picture below shows how far the cup moved after the ball hit it.

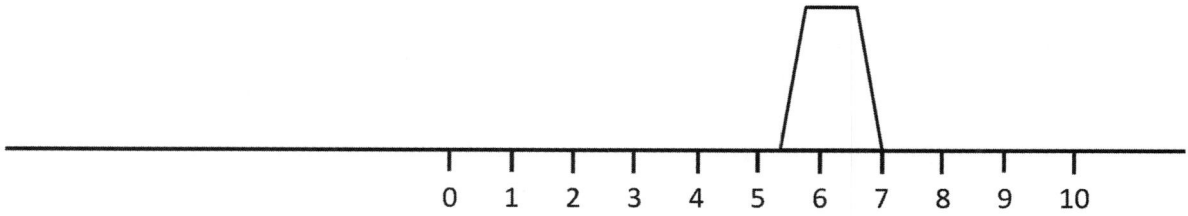

1. How many units did the cup move after the ball hit it?

    a. 4

    b. 5

    c. 6

    d. 7

2. How much would the cup have moved if the ball was moving faster before it hit the cup?

    a. The cup would have moved fewer units.

    b. The cup would have moved more units.

    c. The cup would have moved the same number of units.

Student Workbook for Argument-Driven Inquiry in **Fourth-Grade Science:** Three-Dimensional Investigations

Checkout Questions

3. Explain your thinking. What *cause-and-effect relationship* did you use to answer question 2?

_____

_____

_____

_____

_____

_____

_____

_____

_____

_____

Teacher Scoring Rubric for the Checkout Questions

Level	Description
3	The student can apply the core idea correctly in all cases and can fully explain the cause-and-effect relationship.
2	The student can apply the core idea correctly in all cases but cannot fully explain the cause-and-effect relationship.
1	The student cannot apply the core idea correctly in all cases but can fully explain the cause-and-effect relationship.
0	The student cannot apply the core idea correctly in all cases and cannot explain the cause-and-effect relationship.

Investigation Log

Investigation 2

Energy Transfer Through Collisions: How Does the Energy of a Moving Ball Change After It Collides With Objects That Have Different Masses?

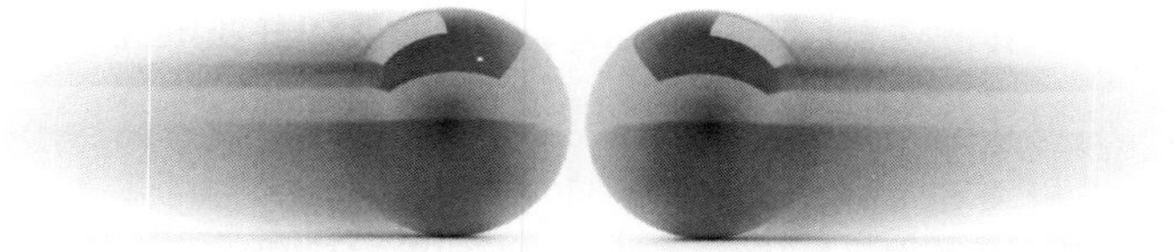

Introduction

When a moving object collides with another object, the motion of the two objects can change in interesting ways. Take a moment to explore what happens when you roll a marble down a ramp and it collides with a second marble. As you explore what happens as a result of the collision, keep track of what you observe and what you are wondering about in the boxes below.

Things I OBSERVED …	Things I WONDER about …

Investigation Log

Energy is transferred between objects during a collision. This transfer of energy causes the motion of these objects to change. You can use this important idea to help explain how the motion of the marbles changed after the collision. When you let go of the marble that you placed on the ramp, it started to move. When the moving marble hit the other marble at the bottom of the ramp, energy transferred from the moving marble to the marble that was not moving. The transfer of energy during the collision caused one marble to slow down and the other to speed up. How much the motion of an object changes as a result of a collision depends on the speed and the masses of the two objects involved in the collision. It is therefore important to understand and to be able predict how the energy of a moving object will change after it collides with objects of different masses.

In this investigation, your goal is to figure out how the energy of a moving billiard ball will change when it collides with objects of different masses. You can create objects of different masses by hanging different numbers of washers from a string in the path of the billiard ball. You can measure the energy of a moving billiard ball by calculating its speed. You can use speed to measure the energy of a moving object because the energy of the moving object increases the faster it moves. Placing the billiard ball at the same starting point on the ramp for each test will cause the ball to travel at the same speed each time before it collides with an object. As you design and carry out your investigation, be sure to keep in mind that energy is always transferred between objects during a collision and that a transfer of energy into or out of an object can change the speed of that object.

Things we KNOW from what we read …	What we will NEED to figure out …

Investigation 2. Energy Transfer Through Collisions:
How Does the Energy of a Moving Ball Change After It Collides
With Objects That Have Different Masses?

Your Task

Use what you know about the transfer of energy, the importance of tracking energy in a system, and cause-and-effect relationships to design and carry out an investigation to examine how a collision with objects of different mass (a *cause*) affects the energy of a moving object (an *effect*).

The *guiding question* of this investigation is, *How does the energy of a moving ball change after it collides with objects that have different masses?*

Materials

You may use any of the following materials during your investigation:

- Safety glasses or goggles (required)
- Wood block (4" × 2" × 2")
- Grooved ruler
- Billiard ball
- Wood stand (8")
- String
- Stopwatch
- 6 washers
- Meterstick
- Scissors

Safety Rules

Follow all normal safety rules. In addition, be sure to follow these rules:

- Wear sanitized safety glasses or goggles during setup, investigation activity, and cleanup.
- Do not throw objects or put any objects in your mouth.
- Keep fingers and toes away from moving objects.
- Be careful when using sharp tools or materials, because they can cut or puncture skin.
- Immediately pick up any items dropped on the floor so they do not become a slip or fall hazard.
- Wash your hands with soap and water when you are done collecting the data.

Plan Your Investigation

Prepare a plan for your investigation by filling out the chart that follows; this plan is called an *investigation proposal*. Before you start developing your plan, be sure to discuss the following questions with the other members of your group:

- What information do we need to find a relationship between a **cause** and an **effect**?
- How can we **track the transfer of energy between objects?**

Our guiding question:

This is a picture of how we will set up the equipment:

We will collect the following data:

These are the steps we will follow to collect data:

I approve of this investigation proposal.

_____ _____
Teacher's signature   Date

National Science Teachers Association

Investigation 2. Energy Transfer Through Collisions: How Does the Energy of a Moving Ball Change After It Collides With Objects That Have Different Masses?

Collect Your Data

Keep a record of what you measure or observe during your investigation in the space below.

Analyze Your Data

You will need to analyze the data you collected before you can develop an answer to the guiding question. To do this, create a graph that shows the relationship between the cause and the effect.

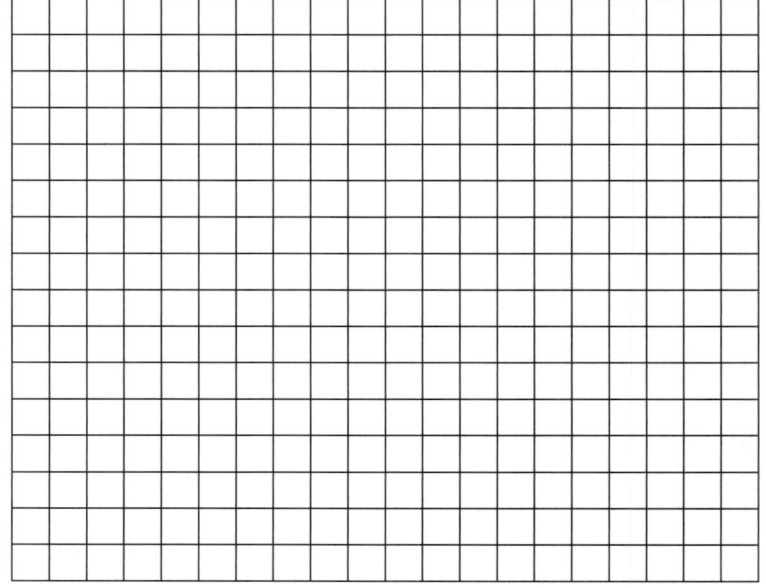

Investigation Log

Draft Argument

Develop an argument on a whiteboard. It should include the following:

1. A *claim:* Your answer to the guiding question.
2. *Evidence:* An analysis of the data and an explanation of what the analysis means.
3. A *justification of the evidence:* Why your group thinks the evidence is important.

The Guiding Question:	
Our Claim:	
Our Evidence:	Our Justification of the Evidence:

Argumentation Session

Share your argument with your classmates. Be sure to ask them how to make your draft argument better. Keep track of their suggestions in the space below.

Ways to IMPROVE our argument …

National Science Teachers Association

Investigation 2. Energy Transfer Through Collisions:
How Does the Energy of a Moving Ball Change After It Collides
With Objects That Have Different Masses?

Draft Report

Prepare an *investigation report* to share what you have learned. Use the information in this handout and your group's final argument to write a *draft* of your investigation report.

Introduction

We have been studying _____ in class.

Before we started this investigation, we explored _____

_____

_____

We noticed _____

_____

_____

My goal for this investigation was to figure out _____

_____

_____

The guiding question was _____

_____

_____

Method

To gather the data I needed to answer this question, I _____

_____

_____

_____

_____

_____

Student Workbook for Argument-Driven Inquiry in **Fourth-Grade Science**: Three-Dimensional Investigations

Investigation Log

I then analyzed the data I collected by _____

_____

_____

_____

Argument

My claim is _____

_____

_____

_____

The graph below shows _____

_____

_____

_____

_____

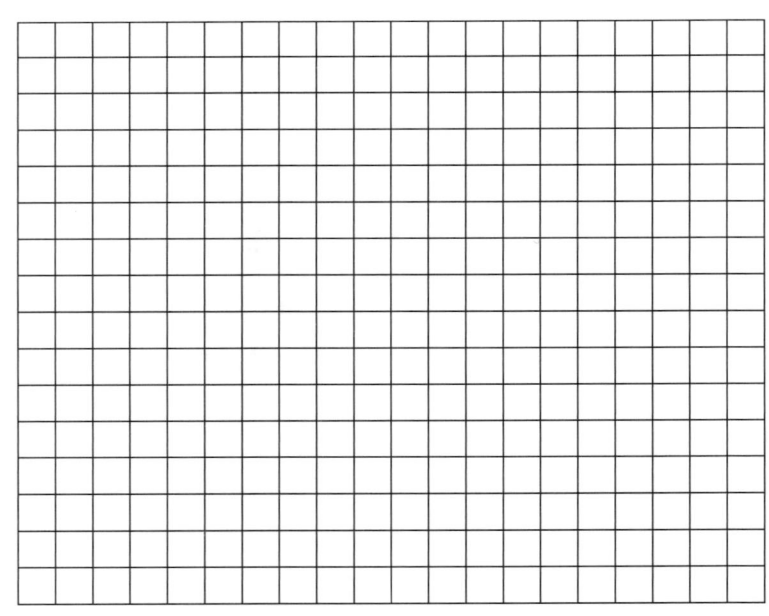

Investigation 2. Energy Transfer Through Collisions: How Does the Energy of a Moving Ball Change After It Collides With Objects That Have Different Masses?

This analysis of the data I collected suggests _____

_____

_____

_____

_____

This evidence is based on several important scientific concepts. The first one is _____

_____

_____

_____

_____

_____

_____

_____

_____

 **Review**

Your friends need your help! Review the draft of their investigation reports and give them ideas about how to improve. Use the *peer-review guide* that begins on the next page to guide your review.

Peer-Review Guide

Section 1: The Investigation	Reviewer Rating		
1. Did the author do a good job of explaining what the investigation was about?	☐ No	☐ Almost	☐ Yes
2. Did the author do a good job of making the **guiding question** clear?	☐ No	☐ Almost	☐ Yes
3. Did the author do a good job of describing what he or she did to **collect data?**	☐ No	☐ Almost	☐ Yes
4. Did the author do a good job describing **how** he or she **analyzed** the data?	☐ No	☐ Almost	☐ Yes
Reviewers: If your group gave the author any "No" or "Almost" ratings, please give the author some advice about what to do to improve this part of his or her investigation report.			

Section 2: The Argument	Reviewer Rating		
1. Does the author's claim provide a clear and detailed **answer** to the guiding question?	☐ No	☐ Almost	☐ Yes
2. Did the author support his or her claim with **scientific evidence?** Scientific evidence includes analyzed data and an explanation of the analysis.	☐ No	☐ Almost	☐ Yes
3. Does the **evidence** that the author uses in his or her argument **support the claim?**	☐ No	☐ Almost	☐ Yes
4. Did the author include enough **evidence** in his or her argument?	☐ No	☐ Almost	☐ Yes
5. Did the author do a good job of **explaining why the evidence** is important (why it matters)?	☐ No	☐ Almost	☐ Yes
6. Is the content of the argument **correct** based on the science concepts we talked about in class?	☐ No	☐ Almost	☐ Yes
Reviewers: If your group gave the author any "No" or "Almost" ratings, please give the author some advice about what to do to improve this part of his or her investigation report.			

Continued

Investigation 2. Energy Transfer Through Collisions:
How Does the Energy of a Moving Ball Change After It Collides
With Objects That Have Different Masses?

Section 3: Mechanics	Reviewer Rating		
1. *Grammar:* Are the sentences complete? Is there proper subject-verb agreement in each sentence? Are there no run-on sentences?	☐ No	☐ Almost	☐ Yes
2. *Conventions:* Did the author use proper spelling, punctuation, and capitalization?	☐ No	☐ Almost	☐ Yes
3. *Word Choice:* Did the author use the right words in each sentence (for example, *there* vs. *their, to* vs. *too, then* vs. *than*)?	☐ No	☐ Almost	☐ Yes

Reviewers: If your group gave the author any "No" or "Almost" ratings, please give the author some advice about what to do to improve the writing mechanics of his or her investigation report.

General Reviewer Comments

We liked …

We wonder …

Investigation Log

Write Your Final Report

Once you have received feedback from your friends about your draft report, create your final investigation report in the space that follows.

Introduction

Method

National Science Teachers Association

Investigation 2. Energy Transfer Through Collisions: How Does the Energy of a Moving Ball Change After It Collides With Objects That Have Different Masses?

Argument

Investigation Report Grading Rubric

Section 1: The Investigation	Missing	Somewhat	Yes
1. The author explained what the investigation was about.	0	1	2
2. The author made the **guiding question** clear.	0	1	2
3. The author **described** what he or she did to **collect data.**	0	1	2
4. The author described **how** he or she **analyzed** the data.	0	1	2

Section 2: The Argument	Missing	Somewhat	Yes
1. The claim includes a clear and detailed **answer** to the guiding question.	0	1	2
2. The author used **scientific evidence** to support the claim. Scientific evidence includes analyzed data and an explanation of the analysis.	0	1	2
3. The evidence **supports the claim.**	0	1	2
4. The author included enough **evidence** in his or her argument.	0	1	2
5. The author **explained why the evidence** is important.	0	1	2
6. The content of the argument is **correct.**	0	1	2

Section 3: Mechanics	Missing	Somewhat	Yes
1. **Grammar:** The sentences are complete. There is proper subject-verb agreement in each sentence. There are no run-on sentences.	0	1	2
2. **Conventions:** The author used proper spelling, punctuation, and capitalization.	0	1	2
3. **Word Choice:** The author used the right words in each sentence (e.g., *there* vs. *their, to* vs. *too, then* vs. *than*).	0	1	2

Teacher Comments

Here are some things I really liked about your report …	Here are some things I think you could do next time to make your report even better …

Total: _____ /26

Checkout Questions

Investigation 2. Energy Transfer Through Collisions

The picture below shows three balls rolling down a ramp. The three balls are moving at the same speed. The balls are about to collide with three different blocks. Each block is the same size but has a different mass.

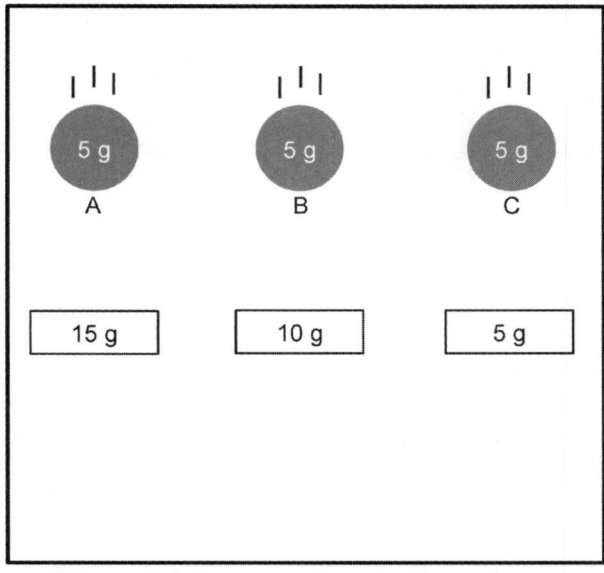

1. Which ball has the *most* energy *before* it collides with the block in front of it?

   a. Ball A

   b. Ball B

   c. Ball C

   d. All three balls have the same amount of energy.

2. Which ball will *slow down the most* after colliding with the block in front of it?

   a. Ball A

   b. Ball B

   c. Ball C

   d. All three balls will be going the same speed.

3. Which ball will have the *most* energy *after* colliding with the block in front of it?

   a. Ball A

   b. Ball B

   c. Ball C

   d. All three balls will have the same amount of energy.

Checkout Questions

4. Explain your thinking. How does the *energy* of a moving ball *change* after a collision?

_____

_____

_____

_____

_____

_____

_____

_____

_____

Teacher Scoring Rubric for the Checkout Questions

Level	Description
3	The student can apply the core idea correctly in all cases and can fully explain the change in energy.
2	The student can apply the core idea correctly in all cases but cannot fully explain the change in energy.
1	The student cannot apply the core idea correctly in all cases but can fully explain the change in energy.
0	The student cannot apply the core idea correctly in all cases and cannot explain the change in energy.

National Science Teachers Association

Investigation Log

Investigation 3

Energy Transfer Between Liquids: How Can We Predict How Much the Temperature of 100 ml of Water Will Change When We Add Different Amounts of Hot Water to It?

Introduction

We use water every day. We use it to cook food, we use it to clean, and we drink it when we are thirsty. Sometimes we need water to be hot, such as when we are cooking, and sometimes we need water to be cold, such as when we drink it with a meal. Take a few minutes to explore what happens to the temperature of some water when you add a hot piece of metal to it. As you explore what happens to the water over time, keep track of what you observe and what you are wondering about in the boxes below.

Things I OBSERVED …	Things I WONDER about …

Investigation Log

All matter, including metal and water, is made up of particles that are in constant motion. These particles have energy because they are constantly moving. The *thermal energy* of an object or substance is the total energy of all the moving particles that make up that object or substance. The *temperature* of an object or substance is a measure of the average energy of these particles.

Energy transfers between objects or substances that are at different temperatures. Energy always transfers from an object or substance at a higher temperature to an object or substance at a lower temperature. For example, when you put the hot piece of metal in the cup of cool water, energy transferred from the hot piece of metal to the cool water; it did not transfer from the cool water to the hot metal. The temperature of the water and the metal changed because energy transferred from the metal into the water. The water increased in temperature because it gained energy from the metal, and the metal decreased in temperature because energy transferred from it to the water.

In this investigation, your goal is to figure out a way to predict how much the temperature of 100 ml of room-temperature water will change when different amounts of hot water are added to it. To accomplish this goal, you will need to find a pattern in the data you collect and then determine a mathematical rule that you can use to describe the pattern. It is useful to be able to predict changes in temperature using mathematics like this because different amounts of hot water will have different amounts of energy even though the hot water is at the same temperature.

Things we KNOW from what we read …	What we will NEED to figure out …

Investigation 3. Energy Transfer Between Liquids: How Can We Predict How Much the Temperature of 100 ml of Water Will Change When We Add Different Amounts of Hot Water to It?

Your Task

Use what you know about the transfer of energy, the importance of tracking energy in a system, and patterns to design and carry out an investigation to determine a rule that you can use to predict how the temperature of 100 ml of room-temperature water will change when different amounts of hot water are added to it.

The *guiding question* of this investigation is, *How can we predict how much the temperature of 100 ml of water will change when we add different amounts of hot water to it?*

Materials

You may use any of the following materials during your investigation:

- Safety goggles (required)
- Nonlatex apron (required)
- Thermal gloves (required)
- 2 beakers (each 250 ml)
- Graduated cylinder (50 ml)
- Thermometer
- Hot water (from your teacher)

Safety Rules

Follow all normal safety rules. In addition, be sure to follow these rules:

- Wear sanitized indirectly vented chemical-splash safety goggles, nonlatex aprons, and thermal gloves during setup, investigation activity, and cleanup.
- Be careful when working with hot water, because it can burn your skin.
- Be careful when working with glassware, which can shatter if dropped and can cut skin.
- Use only GFCI-protected circuits when using electrical equipment, and keep the equipment away from water sources to prevent shock.
- Immediately clean up any spills to avoid a slip or fall hazard.
- Wash your hands with soap and water when you are done collecting the data.

Plan Your Investigation

Prepare a plan for your investigation by filling out the chart that follows; this plan is called an *investigation proposal*. Before you start developing your plan, be sure to discuss the following questions with the other members of your group:

- How can we **track the transfer of energy within a system**?
- What types of **patterns** might we look for to help answer the guiding question?

Investigation Log

Our guiding question:

This is a picture of how we will set up the equipment:

We will collect the following data:

These are the steps we will follow to collect data:

I approve of this investigation proposal.

_____   _____
Teacher's signature   Date

Investigation 3. Energy Transfer Between Liquids: How Can We Predict How Much the Temperature of 100 ml of Water Will Change When We Add Different Amounts of Hot Water to It?

Collect Your Data

Keep a record of what you measure or observe during your investigation in the space below.

Analyze Your Data

You will need to analyze the data you collected before you can develop an answer to the guiding question. To do this, create a graph that shows a pattern or relationship.

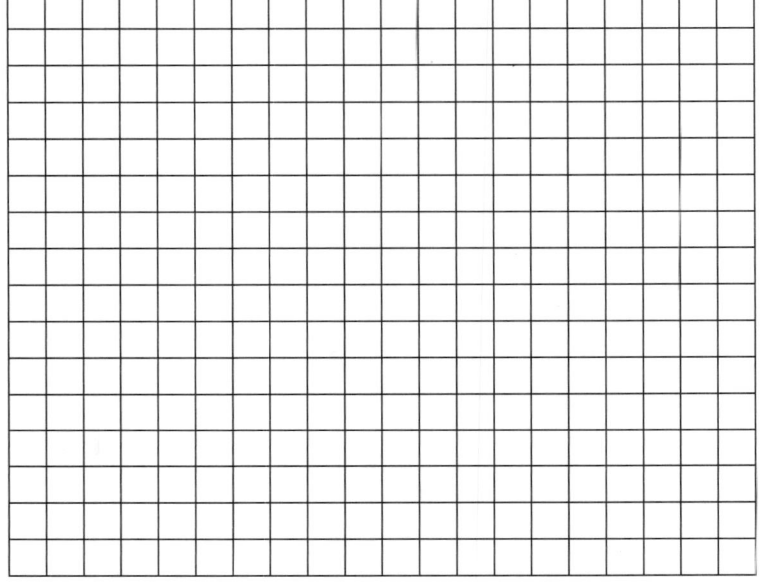

Investigation Log

Draft Argument

Develop an argument on a whiteboard. It should include the following:

1. A *claim*: Your answer to the guiding question.
2. *Evidence*: An analysis of the data and an explanation of what the analysis means.
3. A *justification of the evidence*: Why your group thinks the evidence is important.

The Guiding Question:	
Our Claim:	
Our Evidence:	Our Justification of the Evidence:

Argumentation Session

Share your argument with your classmates. Be sure to ask them how to make your draft argument better. Keep track of their suggestions in the space below.

Ways to IMPROVE our argument …

Investigation 3. Energy Transfer Between Liquids:
How Can We Predict How Much the Temperature of 100 ml of Water Will Change
When We Add Different Amounts of Hot Water to It?

Draft Report

Prepare an *investigation report* to share what you have learned. Use the information in this handout and your group's final argument to write a *draft* of your investigation report.

Introduction

We have been studying _____ in class.

Before we started this investigation, we explored _____

_____

_____

We noticed _____

_____

_____

My goal for this investigation was to figure out _____

_____

_____

The guiding question was _____

_____

_____

Method

To gather the data I needed to answer this question, I _____

_____

_____

_____

_____

Investigation Log

I then analyzed the data I collected by _____

_____

_____

Argument

My claim is _____

_____

_____

_____

_____

The graph below shows _____

_____

_____

_____

_____

_____

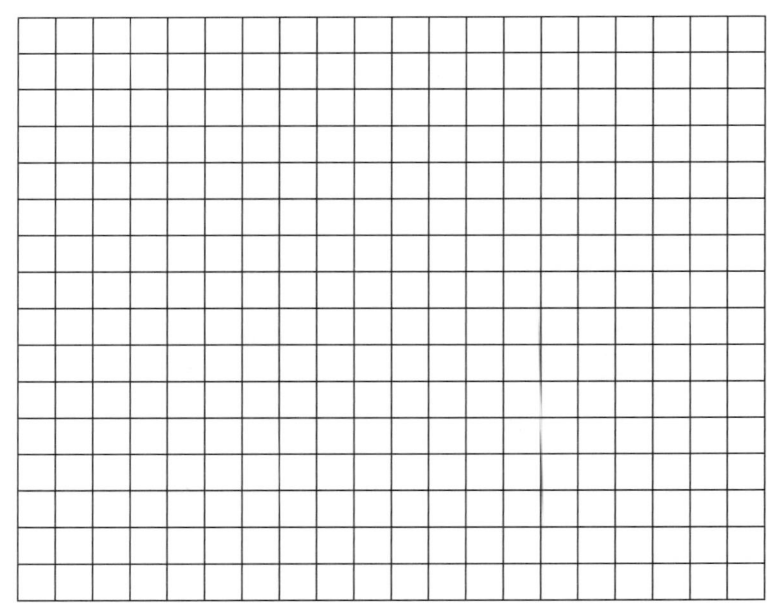

50   National Science Teachers Association

Investigation 3. Energy Transfer Between Liquids: How Can We Predict How Much the Temperature of 100 ml of Water Will Change When We Add Different Amounts of Hot Water to It?

This analysis of the data I collected suggests _____

_____

_____

_____

This evidence is based on several important scientific concepts. The first one is _____

_____

_____

_____

_____

_____

_____

_____

_____

_____

_____

Review

Your friends need your help! Review the draft of their investigation reports and give them ideas about how to improve. Use the *peer-review guide* that begins on the next page to guide your review.

Investigation Log

Peer-Review Guide

Section 1: The Investigation	Reviewer Rating		
1. Did the author do a good job of explaining what the investigation was about?	☐ No	☐ Almost	☐ Yes
2. Did the author do a good job of making the **guiding question** clear?	☐ No	☐ Almost	☐ Yes
3. Did the author do a good job of describing what he or she did to **collect data?**	☐ No	☐ Almost	☐ Yes
4. Did the author do a good job describing **how** he or she **analyzed** the data?	☐ No	☐ Almost	☐ Yes
Reviewers: If your group gave the author any "No" or "Almost" ratings, please give the author some advice about what to do to improve this part of his or her investigation report.			

Section 2: The Argument	Reviewer Rating		
1. Does the author's claim provide a clear and detailed **answer** to the guiding question?	☐ No	☐ Almost	☐ Yes
2. Did the author support his or her claim with **scientific evidence?** Scientific evidence includes analyzed data and an explanation of the analysis.	☐ No	☐ Almost	☐ Yes
3. Does the **evidence** that the author uses in his or her argument **support the claim?**	☐ No	☐ Almost	☐ Yes
4. Did the author include enough **evidence** in his or her argument?	☐ No	☐ Almost	☐ Yes
5. Did the author do a good job of **explaining why the evidence** is important (why it matters)?	☐ No	☐ Almost	☐ Yes
6. Is the content of the argument **correct** based on the science concepts we talked about in class?	☐ No	☐ Almost	☐ Yes
Reviewers: If your group gave the author any "No" or "Almost" ratings, please give the author some advice about what to do to improve this part of his or her investigation report.			

Continued

National Science Teachers Association

Investigation 3. Energy Transfer Between Liquids:
How Can We Predict How Much the Temperature of 100 ml of Water Will Change
When We Add Different Amounts of Hot Water to It?

Section 3: Mechanics	Reviewer Rating		
1. **Grammar:** Are the sentences complete? Is there proper subject-verb agreement in each sentence? Are there no run-on sentences?	☐ No	☐ Almost	☐ Yes
2. **Conventions:** Did the author use proper spelling, punctuation, and capitalization?	☐ No	☐ Almost	☐ Yes
3. **Word Choice:** Did the author use the right words in each sentence (for example, *there* vs. *their, to* vs. *too, then* vs. *than*)?	☐ No	☐ Almost	☐ Yes

Reviewers: If your group gave the author any "No" or "Almost" ratings, please give the author some advice about what to do to improve the writing mechanics of his or her investigation report.

General Reviewer Comments

We liked …

We wonder …

Investigation Log

Write Your Final Report

Once you have received feedback from your friends about your draft report, create your final investigation report in the space that follows.

Introduction

___

Method

___

Investigation 3. Energy Transfer Between Liquids:
How Can We Predict How Much the Temperature of 100 ml of Water Will Change
When We Add Different Amounts of Hot Water to It?

Argument

Investigation Log

Investigation Report Grading Rubric

Section 1: The Investigation	Score Missing	Somewhat	Yes
1. The author explained what the investigation was about.	0	1	2
2. The author made the **guiding question** clear.	0	1	2
3. The author **described** what he or she did to **collect data**.	0	1	2
4. The author described **how** he or she **analyzed** the data.	0	1	2

Section 2: The Argument	Score Missing	Somewhat	Yes
1. The claim includes a clear and detailed **answer** to the guiding question.	0	1	2
2. The author used **scientific evidence** to support the claim. Scientific evidence includes analyzed data and an explanation of the analysis.	0	1	2
3. The evidence **supports the claim.**	0	1	2
4. The author included enough **evidence** in his or her argument.	0	1	2
5. The author **explained why the evidence** is important.	0	1	2
6. The content of the argument is **correct.**	0	1	2

Section 3: Mechanics	Score Missing	Somewhat	Yes
1. *Grammar:* The sentences are complete. There is proper subject-verb agreement in each sentence. There are no run-on sentences.	0	1	2
2. *Conventions:* The author used proper spelling, punctuation, and capitalization.	0	1	2
3. *Word Choice:* The author used the right words in each sentence (e.g., *there* vs. *their, to* vs. *too, then* vs. *than*).	0	1	2

Teacher Comments

Here are some things I really liked about your report …	Here are some things I think you could do next time to make your report even better …

Total: _____ /26

National Science Teachers Association

Checkout Questions

Investigation 3. Energy Transfer Between Liquids

The picture below shows three containers. Each container holds a different amount of room-temperature water. You decide to add 20 ml of 60°C water to each one.

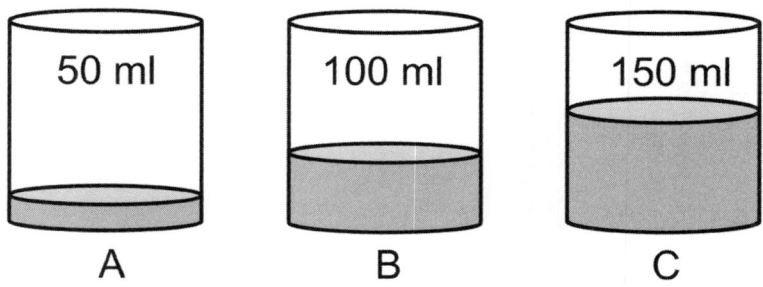

1. Which container of water will increase in temperature the *most*?

    a. Container A
    b. Container B
    c. Container C
    d. They will all increase in temperature by the same amount.

2. Which container of water will increase in temperature the *least*?

    a. Container A
    b. Container B
    c. Container C
    d. They will all increase in temperature by the same amount.

3. Explain your thinking. How does a *transfer of energy* cause a change in temperature?

_____
_____
_____
_____
_____
_____
_____

Teacher Scoring Rubric for the Checkout Questions

Level	Description
3	The student can apply the core idea correctly in all cases and can fully explain the transfer of energy.
2	The student can apply the core idea correctly in all cases but cannot fully explain the transfer of energy.
1	The student cannot apply the core idea correctly in all cases but can fully explain the transfer of energy.
0	The student cannot apply the core idea correctly in all cases and cannot explain the transfer of energy.

Investigation Log

Investigation 4

Energy Transfer by Light: How Does the Distance Between a Light Source and an Object Affect the Amount of Energy an Object Receives From the Light Source?

Introduction

An object in motion can move energy from place to place. Sound and light can also move energy from one place to another. Take a few minutes to explore what happens when you place a thermometer under a piece of black paper and then shine a light on it. As you explore what happens over time, keep track of what you observe and what you are wondering about in the boxes below.

Things I OBSERVED ...	Things I WONDER about ...

Investigation Log

Light can transfer energy from place to place. There are many different types of light. Some light we can see, such as visible light, and some light is invisible to us, such as infrared light or ultraviolet light. Only objects that are at a very high temperature produce light that we can see. For example, the filament in the lightbulb you used earlier got very hot when you turned it on. The filament was so hot that it produced visible light. This light traveled out from the filament in all directions. Some of the light traveled to the black paper. The black paper absorbed the energy that was transferred to it by the light. The transfer of energy from the filament in the lightbulb to the black paper by light caused the temperature of the black paper to increase over time.

Your goal in this investigation is to figure out how the amount of energy an object receives by light changes based on how far it is from or how close it is to a light source. The light source you will use is a lightbulb. You can measure the amount of energy received from a light source by keeping track of how the temperature of an object changes over time. You can then place objects at different distances from a light source. You can examine how the distance between a light source and an object affects the amount of energy that an object gains over time using this method because energy can be transferred from one object to another through light and objects increase in temperature when they gain energy.

Things we KNOW from what we read …	What we will NEED to figure out …

Investigation 4. Energy Transfer by Light:
How Does the Distance Between a Light Source and an Object Affect
the Amount of Energy an Object Receives From the Light Source?

Your Task

Use what you know about energy, light, measurement scales, and the importance of tracking the movement of energy in a system to design and carry out an investigation to learn more about how the amount of energy an object receives by light changes based on how far it is from or how close it is to a light source.

The *guiding question* of this investigation is, ***How does the distance between a light source and an object affect the amount of energy an object receives from the light source?***

Materials

You may use any of the following materials during your investigation:

- Safety glasses or goggles (required)
- Heat lamp
- 3 black mailing tubes (each 2" x 3")
- 3 thermometers
- Cloth meterstick

Safety Rules

Follow all normal safety rules. In addition, be sure to follow these rules:

- Wear sanitized safety glasses or goggles during setup, investigation activity, and cleanup.
- Use only GFCI-protected circuits when using electrical equipment, and keep the equipment away from water sources to prevent shock.
- Be careful when handling the heat lamp and bulb. The bulb can shatter if dropped and can cut skin. Do not touch the bulb when it is on or for several minutes after turning it off, because lightbulbs can get very hot and burn skin.
- Wash your hands with soap and water when you are done collecting the data.

Plan Your Investigation

Prepare a plan for your investigation by filling out the chart that follows; this plan is called an *investigation proposal*. Before you start developing your plan, be sure to discuss the following questions with the other members of your group:

- How can we **track the transfer of energy within a system**?
- How can we track changes by taking **measurements,** and what **scale** should we use?

Investigation Log

Our guiding question:

This is a picture of how we will set up the equipment:

We will collect the following data:

These are the steps we will follow to collect data:

I approve of this investigation proposal.

_____   _____
Teacher's signature   Date

Investigation 4. Energy Transfer by Light:
How Does the Distance Between a Light Source and an Object Affect
the Amount of Energy an Object Receives From the Light Source?

Collect Your Data

Keep a record of what you measure or observe during your investigation in the space below.

Analyze Your Data

You will need to analyze the data you collected before you can develop an answer to the guiding question. To do this, create a graph that shows the relationship between what you changed and what you measured or observed as a result of what you changed.

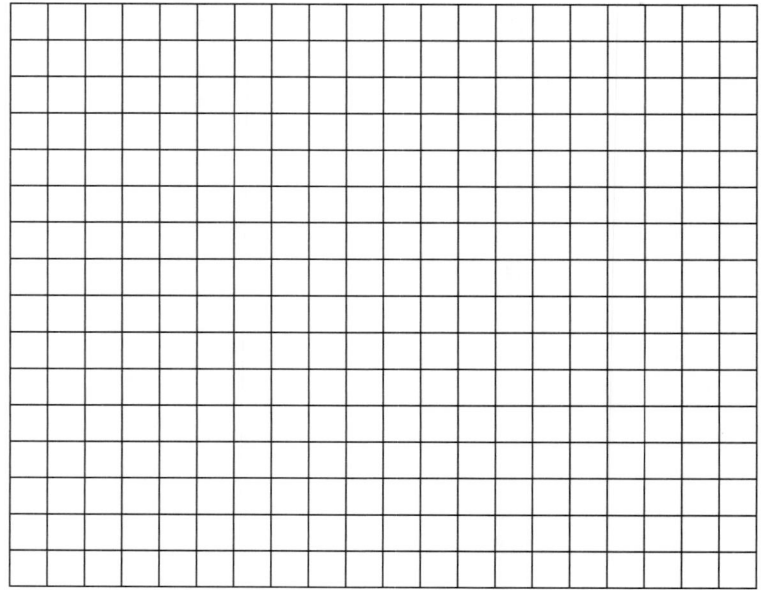

Investigation Log

Draft Argument

Develop an argument on a whiteboard. It should include the following:

1. A *claim*: Your answer to the guiding question.
2. *Evidence*: An analysis of the data and an explanation of what the analysis means.
3. A *justification of the evidence*: Why your group thinks the evidence is important.

The Guiding Question:	
Our Claim:	
Our Evidence:	Our Justification of the Evidence:

Argumentation Session

Share your argument with your classmates. Be sure to ask them how to make your draft argument better. Keep track of their suggestions in the space below.

Ways to IMPROVE our argument …

Investigation 4. Energy Transfer by Light:
How Does the Distance Between a Light Source and an Object Affect
the Amount of Energy an Object Receives From the Light Source?

Draft Report

Prepare an *investigation report* to share what you have learned. Use the information in this handout and your group's final argument to write a *draft* of your investigation report.

Introduction

We have been studying _____ in class.

Before we started this investigation, we explored _____
_____
_____

We noticed _____
_____
_____

My goal for this investigation was to figure out _____
_____
_____

The guiding question was _____
_____
_____
_____

Method

To gather the data I needed to answer this question, I _____
_____
_____
_____
_____

Investigation Log

I then analyzed the data I collected by _____

_____

_____

_____

_____

_____

Argument

My claim is _____

_____

_____

_____

The graph below shows _____

_____

_____

_____

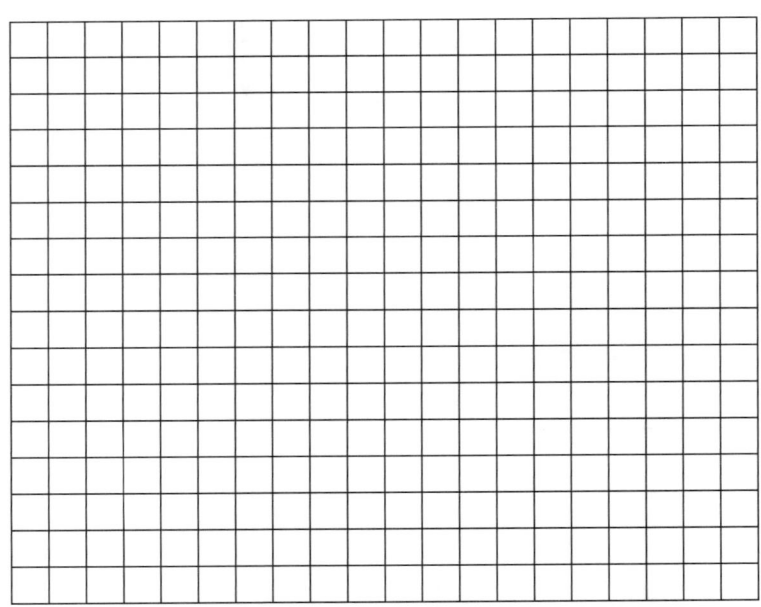

Investigation 4. Energy Transfer by Light: How Does the Distance Between a Light Source and an Object Affect the Amount of Energy an Object Receives From the Light Source?

This analysis of the data I collected suggests _____

_____

_____

_____

_____

This evidence is important because of several different scientific concepts. The first one is _____

_____

_____

_____

_____

_____

_____

_____

_____

_____

 **Review**

Your friends need your help! Review the draft of their investigation reports and give them ideas about how to improve. Use the *peer-review guide* that begins on the next page to guide your review.

Investigation Log

Peer-Review Guide

Section 1: The Investigation	Reviewer Rating		
1. Did the author do a good job of explaining what the investigation was about?	☐ No	☐ Almost	☐ Yes
2. Did the author do a good job of making the **guiding question** clear?	☐ No	☐ Almost	☐ Yes
3. Did the author do a good job of describing what he or she did to **collect data?**	☐ No	☐ Almost	☐ Yes
4. Did the author do a good job describing **how** he or she **analyzed** the data?	☐ No	☐ Almost	☐ Yes

Reviewers: If your group gave the author any "No" or "Almost" ratings, please give the author some advice about what to do to improve this part of his or her investigation report.

Section 2: The Argument	Reviewer Rating		
1. Does the author's claim provide a clear and detailed **answer** to the guiding question?	☐ No	☐ Almost	☐ Yes
2. Did the author support his or her claim with **scientific evidence?** Scientific evidence includes analyzed data and an explanation of the analysis.	☐ No	☐ Almost	☐ Yes
3. Does the **evidence** that the author uses in his or her argument **support the claim?**	☐ No	☐ Almost	☐ Yes
4. Did the author include enough **evidence** in his or her argument?	☐ No	☐ Almost	☐ Yes
5. Did the author do a good job of **explaining why the evidence** is important (why it matters)?	☐ No	☐ Almost	☐ Yes
6. Is the content of the argument **correct** based on the science concepts we talked about in class?	☐ No	☐ Almost	☐ Yes

Reviewers: If your group gave the author any "No" or "Almost" ratings, please give the author some advice about what to do to improve this part of his or her investigation report.

Continued

Investigation 4. Energy Transfer by Light:
How Does the Distance Between a Light Source and an Object Affect
the Amount of Energy an Object Receives From the Light Source?

Section 3: Mechanics	Reviewer Rating		
1. *Grammar:* Are the sentences complete? Is there proper subject-verb agreement in each sentence? Are there no run-on sentences?	☐ No	☐ Almost	☐ Yes
2. *Conventions:* Did the author use proper spelling, punctuation, and capitalization?	☐ No	☐ Almost	☐ Yes
3. *Word Choice:* Did the author use the right words in each sentence (for example, *there* vs. *their, to* vs. *too, then* vs. *than*)?	☐ No	☐ Almost	☐ Yes

Reviewers: If your group gave the author any "No" or "Almost" ratings, please give the author some advice about what to do to improve the writing mechanics of his or her investigation report.

General Reviewer Comments

We liked …

We wonder …

Investigation Log

Write Your Final Report

Once you have received feedback from your friends about your draft report, create your final investigation report in the space that follows.

Introduction

Method

Investigation 4. Energy Transfer by Light:
How Does the Distance Between a Light Source and an Object Affect
the Amount of Energy an Object Receives From the Light Source?

Argument

_____
_____
_____
_____
_____
_____
_____
_____
_____

_____
_____
_____
_____
_____
_____

Investigation Log

Investigation Report Grading Rubric

Section 1: The Investigation	Score Missing	Somewhat	Yes
1. The author explained what the investigation was about.	0	1	2
2. The author made the **guiding question** clear.	0	1	2
3. The author **described** what he or she did to **collect data.**	0	1	2
4. The author described **how** he or she **analyzed** the data.	0	1	2

Section 2: The Argument	Score Missing	Somewhat	Yes
1. The claim includes a clear and detailed **answer** to the guiding question.	0	1	2
2. The author used **scientific evidence** to support the claim. Scientific evidence includes analyzed data and an explanation of the analysis.	0	1	2
3. The evidence **supports the claim.**	0	1	2
4. The author included enough **evidence** in his or her argument.	0	1	2
5. The author **explained why the evidence** is important.	0	1	2
6. The content of the argument is **correct.**	0	1	2

Section 3: Mechanics	Score Missing	Somewhat	Yes
1. *Grammar:* The sentences are complete. There is proper subject-verb agreement in each sentence. There are no run-on sentences.	0	1	2
2. *Conventions:* The author used proper spelling, punctuation, and capitalization.	0	1	2
3. *Word Choice:* The author used the right words in each sentence (e.g., *there* vs. *their, to* vs. *too, then* vs. *than*).	0	1	2

Teacher Comments

Here are some things I really liked about your report …	Here are some things I think you could do next time to make your report even better …

Total: _____ /26

Checkout Questions

Investigation 4. Energy Transfer by Light

The picture at right shows three cubes sitting on a table. Each cube is made of the same type of metal.

The cubes are sitting on the table at different distances from a heat lamp. The lamp has been on for 10 minutes.

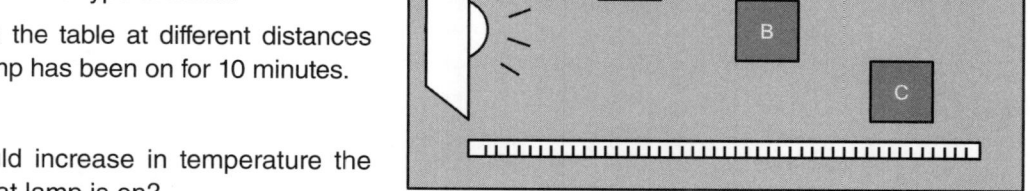

1. Which cube should increase in temperature the *most* while the heat lamp is on?

    a. Cube A

    b. Cube B

    c. Cube C

    d. They will all increase in temperature by the same amount.

2. Which cube should increase in temperature the *least* while the heat lamp is on?

    a. Cube A

    b. Cube B

    c. Cube C

    d. They will all increase in temperature by the same amount.

3. Explain your thinking. How does a *transfer of energy* cause a change in temperature?

_____

_____

_____

_____

_____

_____

Teacher Scoring Rubric for the Checkout Questions

Level	Description
3	The student can apply the core idea correctly in all cases and can fully explain the transfer of energy.
2	The student can apply the core idea correctly in all cases but cannot fully explain the transfer of energy.
1	The student cannot apply the core idea correctly in all cases but can fully explain the transfer of energy.
0	The student cannot apply the core idea correctly in all cases and cannot explain the transfer of energy.

Investigation Log

Investigation 5

Electric Currents: How Does Adding More Batteries or Bulbs to a Closed Circuit Affect the Brightness of a Single Bulb Within That Circuit?

Introduction

We use lightbulbs to create the light we need to be able to see when it is dark. Take a few minutes to see if you can find four different ways to make a lightbulb light up using only one battery and one piece of wire. As you connect the wire, battery, and bulb together in different ways, keep track of what you observe and what you are wondering about in the boxes below.

Things I OBSERVED …	Things I WONDER about …

Student Workbook for Argument-Driven Inquiry in **Fourth-Grade Science**: Three-Dimensional Investigations

Investigation Log

The bulb lights up when you create a *closed circuit* with the wire and the battery. A closed circuit is a path or loop that connects the positive and negative ends of a battery together. The battery creates an *electric current* that moves through a closed circuit. The electric current moves in one direction. It moves from the positive terminal of the battery toward the negative terminal of the battery. As the electric current travels through the bulb in the circuit, it must move through a very small wire called the *filament*. The filament heats up as the electric current flows through it. The filament gets so hot that it gives off visible light. Over time, all of the energy stored in a battery that is used to create the electric current is transformed into heat and light inside the bulb.

There are many ways to create a closed circuit. For example, you can make a closed circuit that includes one battery and one or more different bulbs. You can also make a closed circuit with one bulb and one or more different batteries. Your goal in this investigation is to figure out how the addition of more batteries and bulbs to a closed circuit affects the brightness of a bulb in that circuit. You will need to think about how to create a closed circuit with more than one battery or bulb in it. Your teacher will tell you how to measure the brightness of the bulb. As you design your investigation, be sure to keep in mind that electric currents can transfer energy from place to place, electric currents move through a closed circuit, batteries are used to create electric currents, and an electric current moving through a filament of a bulb creates heat and light.

Things we KNOW from what we read …	What we will NEED to figure out …

Investigation 5. Electric Currents:
How Does Adding More Batteries or Bulbs to a Closed Circuit Affect
the Brightness of a Single Bulb Within That Circuit?

Your Task

Use what you know about electric currents, closed circuits, the importance of tracking the movement of energy in a system, and measurement scales to design and carry out an investigation to learn more how the addition of more batteries and bulbs to a closed circuit affects the brightness of a single bulb in that circuit.

The *guiding question* of this investigation is, **How does adding more batteries or bulbs to a closed circuit affect the brightness of a single bulb within that circuit?**

Materials

You may use any of the following materials during your investigation:

- Safety goggles (required)
- 3 miniature lightbulbs
- 3 mini bulb holders
- 20 small pieces of paper
- 6 double-ended alligator clips to test lead wire
- 3 AA batteries
- 3 battery holders

Safety Rules

Follow all normal safety rules. In addition, be sure to follow these rules:

- Wear sanitized indirectly vented chemical-splash goggles during setup, investigation activity, and cleanup.
- Be careful when using sharp tools or materials, because they can cut or puncture skin.
- Be careful when handling lightbulbs, which are made of glass and can shatter and cut skin. If they break, do not touch the glass and tell your teacher as soon as you can. Do not touch a lightbulb when it is on or for several minutes after turning it off, because lightbulbs can get very hot and burn skin.
- Wash your hands with soap and water when you are done collecting the data.

Plan Your Investigation

Prepare a plan for your investigation by filling out the chart that follows; this plan is called an *investigation proposal*. Before you start developing your plan, be sure to discuss the following questions with the other members of your group:

- How might the **parts** of a circuit be related to or affect how it **functions**?
- How can we track changes by taking **measurements**, and what **scale** should we use?

Investigation Log

Our guiding question:

This is a picture of how we will set up the equipment:

We will collect the following data:

These are the steps we will follow to collect data:

I approve of this investigation proposal.

_____ _____
Teacher's signature   Date

National Science Teachers Association

Investigation 5. Electric Currents:
How Does Adding More Batteries or Bulbs to a Closed Circuit Affect
the Brightness of a Single Bulb Within That Circuit?

Collect Your Data

Keep a record of what you measure or observe during your investigation in the space below.

Analyze Your Data

You will need to analyze the data you collected before you can develop an answer to the guiding question. To do this, create two graphs. Each graph should show the relationship between what you changed and what you measured or observed as a result of what you changed.

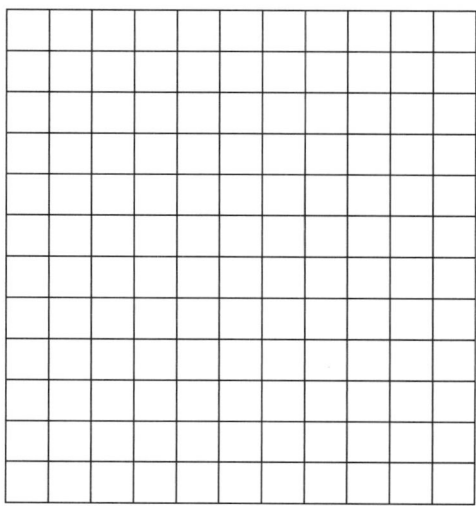
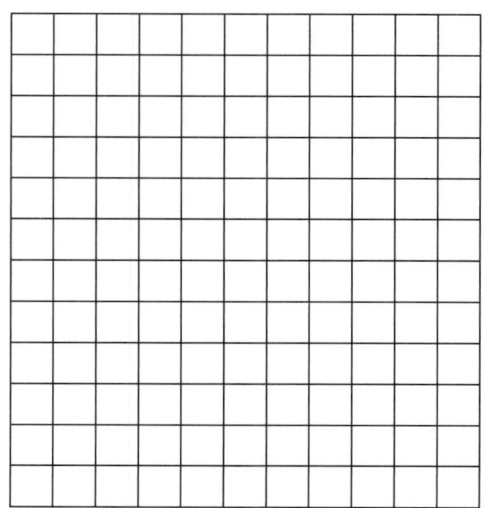

Student Workbook for Argument-Driven Inquiry in **Fourth-Grade Science**: Three-Dimensional Investigations

Investigation Log

Draft Argument

Develop an argument on a whiteboard. It should include the following:

1. A *claim*: Your answer to the guiding question.
2. *Evidence*: An analysis of the data and an explanation of what the analysis means.
3. A *justification of the evidence*: Why your group thinks the evidence is important.

The Guiding Question:	
Our Claim:	
Our Evidence:	Our Justification of the Evidence:

Argumentation Session

Share your argument with your classmates. Be sure to ask them how to make your draft argument better. Keep track of their suggestions in the space below.

Ways to IMPROVE our argument …

National Science Teachers Association

Investigation 5. Electric Currents:
How Does Adding More Batteries or Bulbs to a Closed Circuit Affect
the Brightness of a Single Bulb Within That Circuit?

Draft Report

Prepare an investigation report to share what you have learned. Use the information in this handout and your group's final argument to write a draft of your investigation report.

Introduction

We have been studying _____ in class.

Before we started this investigation, we explored _____

_____

_____

We noticed _____

_____

_____

My goal for this investigation was to figure out _____

_____

_____

_____

The guiding question was _____

_____

_____

Method

To gather the data I needed to answer this question, I _____

_____

_____

_____

_____

Investigation Log

I then analyzed the data I collected by _____
_____
_____
_____

Argument

My claim is _____
_____
_____

The graphs below show _____
_____
_____
_____

This analysis of the data I collected suggests _____
_____
_____

Investigation 5. Electric Currents:
How Does Adding More Batteries or Bulbs to a Closed Circuit Affect
the Brightness of a Single Bulb Within That Circuit?

_____
_____
_____
_____

This evidence is important because of several scientific concepts. The first one is _____

_____
_____
_____
_____
_____
_____
_____
_____
_____
_____

Review

Your friends need your help! Review the draft of their investigation reports and give them ideas about how to improve. Use the *peer-review guide* that begins on the next page to guide your review.

Peer-Review Guide

Section 1: The Investigation	Reviewer Rating		
1. Did the author do a good job of explaining what the investigation was about?	☐ No	☐ Almost	☐ Yes
2. Did the author do a good job of making the **guiding question** clear?	☐ No	☐ Almost	☐ Yes
3. Did the author do a good job of describing what he or she did to **collect data**?	☐ No	☐ Almost	☐ Yes
4. Did the author do a good job describing **how** he or she **analyzed** the data?	☐ No	☐ Almost	☐ Yes
Reviewers: If your group gave the author any "No" or "Almost" ratings, please give the author some advice about what to do to improve this part of his or her investigation report.			

Section 2: The Argument	Reviewer Rating		
1. Does the author's claim provide a clear and detailed **answer** to the guiding question?	☐ No	☐ Almost	☐ Yes
2. Did the author support his or her claim with **scientific evidence**? Scientific evidence includes analyzed data and an explanation of the analysis.	☐ No	☐ Almost	☐ Yes
3. Does the **evidence** that the author uses in his or her argument **support the claim**?	☐ No	☐ Almost	☐ Yes
4. Did the author include enough **evidence** in his or her argument?	☐ No	☐ Almost	☐ Yes
5. Did the author do a good job of **explaining why the evidence** is important (why it matters)?	☐ No	☐ Almost	☐ Yes
6. Is the content of the argument **correct** based on the science concepts we talked about in class?	☐ No	☐ Almost	☐ Yes
Reviewers: If your group gave the author any "No" or "Almost" ratings, please give the author some advice about what to do to improve this part of his or her investigation report.			

Continued

Investigation 5. Electric Currents:
How Does Adding More Batteries or Bulbs to a Closed Circuit Affect
the Brightness of a Single Bulb Within That Circuit?

Section 3: Mechanics	Reviewer Rating		
1. *Grammar:* Are the sentences complete? Is there proper subject-verb agreement in each sentence? Are there no run-on sentences?	☐ No	☐ Almost	☐ Yes
2. *Conventions:* Did the author use proper spelling, punctuation, and capitalization?	☐ No	☐ Almost	☐ Yes
3. *Word Choice:* Did the author use the right words in each sentence (for example, *there* vs. *their, to* vs. *too, then* vs. *than*)?	☐ No	☐ Almost	☐ Yes

Reviewers: If your group gave the author any "No" or "Almost" ratings, please give the author some advice about what to do to improve the writing mechanics of his or her investigation report.

General Reviewer Comments

We liked …

We wonder …

Investigation Log

Write Your Final Report

Once you have received feedback from your friends about your draft report, create your final investigation report in the space that follows.

Introduction

_____
_____
_____
_____
_____
_____
_____

Method

_____
_____
_____
_____
_____
_____
_____
_____
_____
_____
_____

Investigation 5. Electric Currents:
How Does Adding More Batteries or Bulbs to a Closed Circuit Affect
the Brightness of a Single Bulb Within That Circuit?

Argument

Investigation Log

Investigation Report Grading Rubric

Section 1: The Investigation	Missing	Somewhat	Yes
		Score	
1. The author explained what the investigation was about.	0	1	2
2. The author made the **guiding question** clear.	0	1	2
3. The author **described** what he or she did to **collect data.**	0	1	2
4. The author described **how** he or she **analyzed** the data.	0	1	2

Section 2: The Argument	Missing	Somewhat	Yes
		Score	
1. The claim includes a clear and detailed **answer** to the guiding question.	0	1	2
2. The author used **scientific evidence** to support the claim. Scientific evidence includes analyzed data and an explanation of the analysis.	0	1	2
3. The evidence **supports the claim.**	0	1	2
4. The author included enough **evidence** in his or her argument.	0	1	2
5. The author **explained why the evidence** is important.	0	1	2
6. The content of the argument is **correct.**	0	1	2

Section 3: Mechanics	Missing	Somewhat	Yes
		Score	
1. *Grammar:* The sentences are complete. There is proper subject-verb agreement in each sentence. There are no run-on sentences.	0	1	2
2. *Conventions:* The author used proper spelling, punctuation, and capitalization.	0	1	2
3. *Word Choice:* The author used the right words in each sentence (e.g., *there* vs. *their, to* vs. *too, then* vs. *than*).	0	1	2

Teacher Comments

Here are some things I really liked about your report …	Here are some things I think you could do next time to make your report even better …

Total: _____ /26

Checkout Questions

Investigation 5. Electric Currents

1. The pictures below show three different ways to connect a bulb to a battery using a wire.

 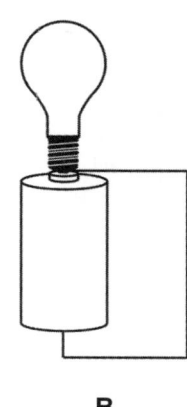 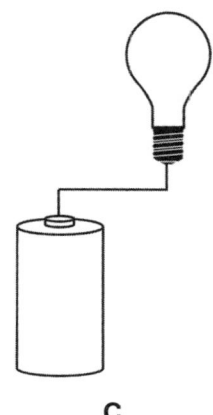

            A   B   C

Which way would make the bulb light up?

    a. A only

    b. B only

    c. C only

    d. A and B only

    e. B and C only

    f. A and C only

    g. A, B, and C

2. The pictures below show three different circuits.

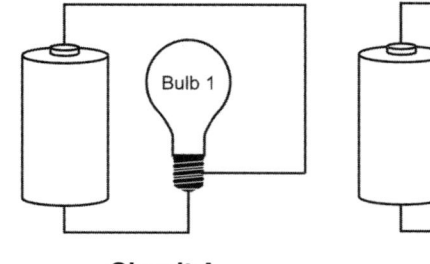 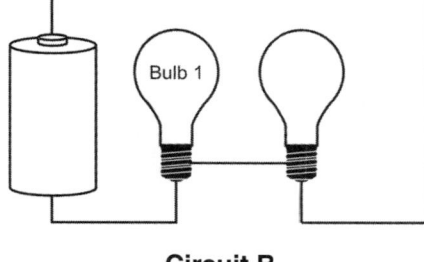 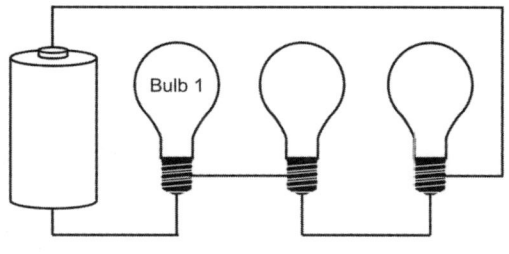

    Circuit A   **Circuit B**   **Circuit C**

In which circuit would bulb 1 be the brightest?

    a. Circuit A

    b. Circuit B

    c. Circuit C

    d. Bulb 1 will be equally bright in all three circuits.

Checkout Questions

3. The pictures below show three different circuits.

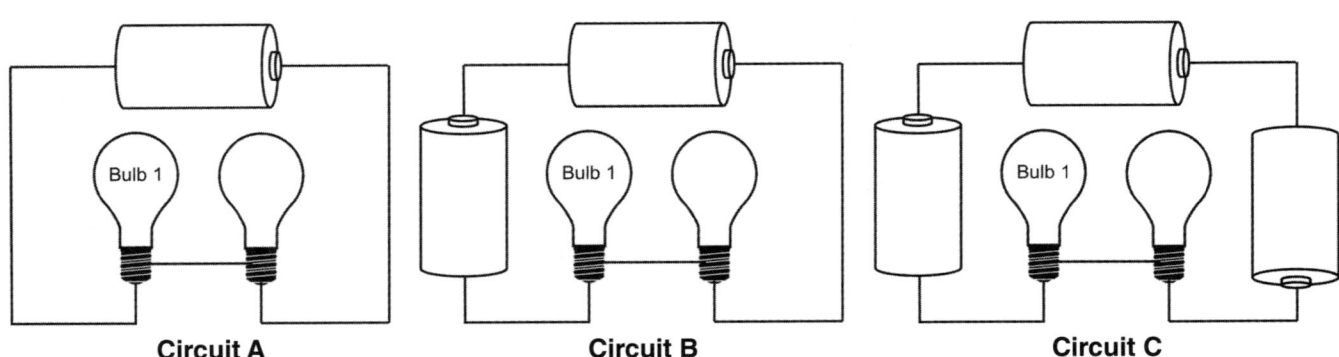

Circuit A   Circuit B   Circuit C

In which circuit would bulb 1 be the brightest?

   a. Circuit A
   b. Circuit B
   c. Circuit C
   d. Bulb 1 will be equally bright in all three circuits.

4. Explain your thinking. How does what you know about the *relationship between structure and function* help you make these predictions?

_____
_____
_____
_____
_____
_____
_____
_____

Teacher Scoring Rubric for the Checkout Questions

Level	Description
3	The student can apply the core idea correctly in all cases and can fully explain how structure and function are related.
2	The student can apply the core idea correctly in all cases but cannot fully explain how structure and function are related.
1	The student cannot apply the core idea correctly in all cases but can fully explain how structure and function are related.
0	The student cannot apply the core idea correctly in all cases and cannot explain how structure and function are related.

National Science Teachers Association

Investigation Log

Investigation 6

Energy Transfer by Electric Currents: How Can We Make an Electric Car Move Faster?

Introduction

People can convert electrical energy into motion by adding a motor to a closed circuit. Take a few minutes to see if you can make a motor start to move using a battery and two wires. Be sure to keep track of what you observe and what you are wondering about in the boxes below.

Things I OBSERVED ...	Things I WONDER about ...

Investigation Log

The motor started to move when you created a *closed circuit* with the wire and the battery. A closed circuit is a path or loop that connects the positive and negative ends of a battery. The battery creates an *electric current* that moves through a closed circuit. The electric current moves in one direction. It moves from the positive terminal of the battery toward the negative terminal of the battery. An *electric motor* is a device that converts an electric current into physical movement. People often use electric motors to turn wheels, propellers, or winches. Over time, all of the energy stored in a battery that is used to create the electric current is transformed into heat or movement. This transfer of energy from one form to another form is why we need to replace or recharge batteries after using them to make an electric motor run.

Your goal in this investigation is to apply what you know about electric currents, closed circuits, and the transfer of energy to create an electric car. You then need to figure out how to make the car move faster by changing the number of batteries that you use to make the electric motor run. You can measure the speed of your electric car by timing how long it takes for your car to travel a track that is 3 meters long. You can then divide the distance the car traveled (3 meters) by how long it took the car to travel that distance (in seconds). The quotient of this division problem is the speed of your car in meters per second (m/s). As you design your investigation, be sure to keep in mind that batteries store energy, an electric current is movement of electrical energy from one place to another, and a motor converts electrical energy into motion.

Things we KNOW from what we read …	What we will NEED to do and figure out …

Investigation 6. Energy Transfer by Electric Currents: How Can We Make an Electric Car Move Faster?

Your Task

Use what you know about electric currents, closed circuits, the relationship between structure and function, and the importance of tracking the movement of energy in a system to design, test, and refine a car that is able to convert electrical energy into motion.

The *guiding question* of this investigation is, *How can we make an electric car move faster?*

Materials

You may use any of the following materials during your investigation:

- Safety goggles (required)
- Car frame
- Electric motor with propeller
- Rubber bands
- AA batteries
- AA battery holder for one battery
- AA battery holder for two batteries, with switch
- AA battery holder for three batteries, with switch
- AA battery holder for four batteries, with switch
- Stopwatch

Safety Rules

Follow all normal safety rules. In addition, be sure to follow these rules:

- Wear sanitized indirectly vented chemical-splash goggles during setup, investigation activity, and cleanup.
- Do not touch the propellers while they are moving, because they can move very fast and cause injury.
- Keep fingers and toes away from moving objects.
- Be careful when handling sharp objects such as nails and wires, because they can cut or puncture skin.
- Wash your hands with soap and water when you are done collecting the data.

Plan Your Investigation

Prepare a plan for your investigation by filling out the chart that follows; this plan is called an *investigation proposal*. Before you start developing your plan, be sure to discuss the following questions with the other members of your group:

- How can we track the **transfer of energy within a system**?
- How might the **structure** of a car be related to the way the car **functions**?

Investigation Log

Our guiding question:

This is a picture of how we will set up the equipment:

We will collect the following data:

These are the steps we will follow to collect data:

I approve of this investigation proposal.

Teacher's signature

Date

Investigation 6. Energy Transfer by Electric Currents: How Can We Make an Electric Car Move Faster?

Collect Your Data

Keep a record of what you measure or observe during your investigation in the space below.

Analyze Your Data

You will need to analyze the data you collected before you can develop an answer to the guiding question. To do this, create a graph that shows the relationship between what you changed and what you measured during your investigation.

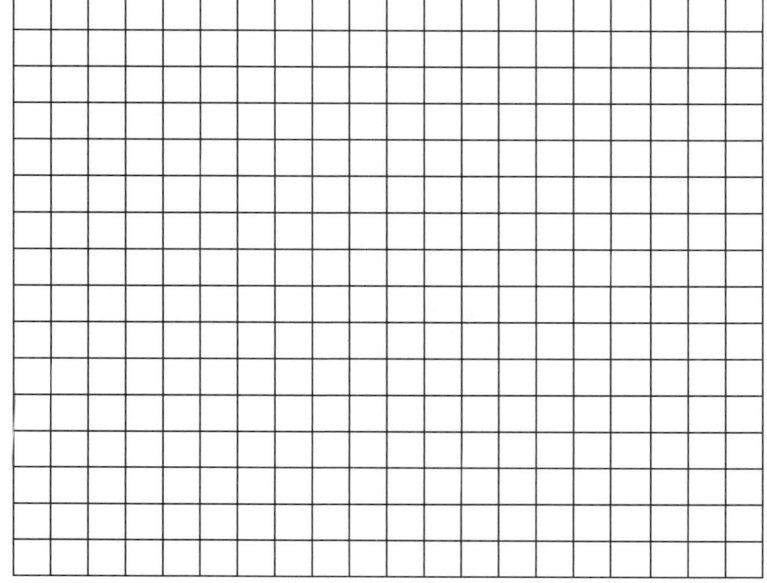

Investigation Log

Draft Argument

Develop an argument on a whiteboard. It should include the following:

1. A *claim*: Your answer to the guiding question.
2. *Evidence*: An analysis of the data and an explanation of what the analysis means.
3. A *justification of the evidence*: Why your group thinks the evidence is important.

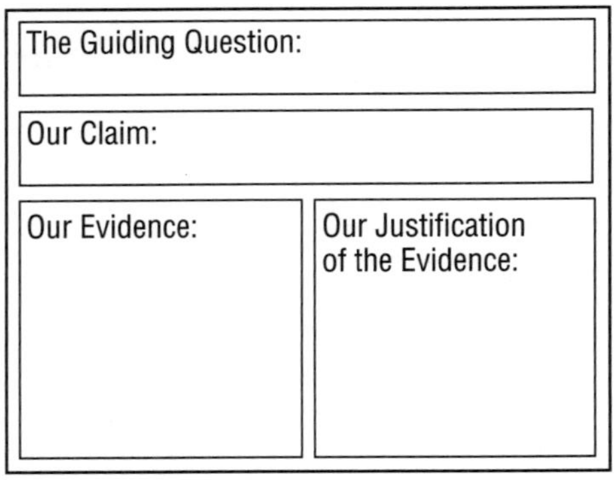

The Guiding Question:	
Our Claim:	
Our Evidence:	Our Justification of the Evidence:

Argumentation Session

Share your argument with your classmates. Be sure to ask them how to make your draft argument better. Keep track of their suggestions in the space below.

Ways to IMPROVE our argument …

Investigation 6. Energy Transfer by Electric Currents: How Can We Make an Electric Car Move Faster?

Draft Report

Prepare an *investigation report* to share what you have learned. Use the information in this handout and your group's final argument to write a *draft* of your investigation report.

Introduction

We have been studying _____ in class.

Before we started this investigation, we explored _____
_____
_____

We noticed _____
_____
_____

My goal for this investigation was to figure out _____
_____
_____

The guiding question was _____
_____
_____

Method

To gather the data I needed to answer this question, I _____
_____
_____
_____
_____

Student Workbook for Argument-Driven Inquiry in **Fourth-Grade Science**: Three-Dimensional Investigations

Investigation Log

I then analyzed the data I collected by _____

_____

_____

_____

Argument

My claim is _____

_____

_____

_____

The graph below shows _____

_____

_____

_____

_____

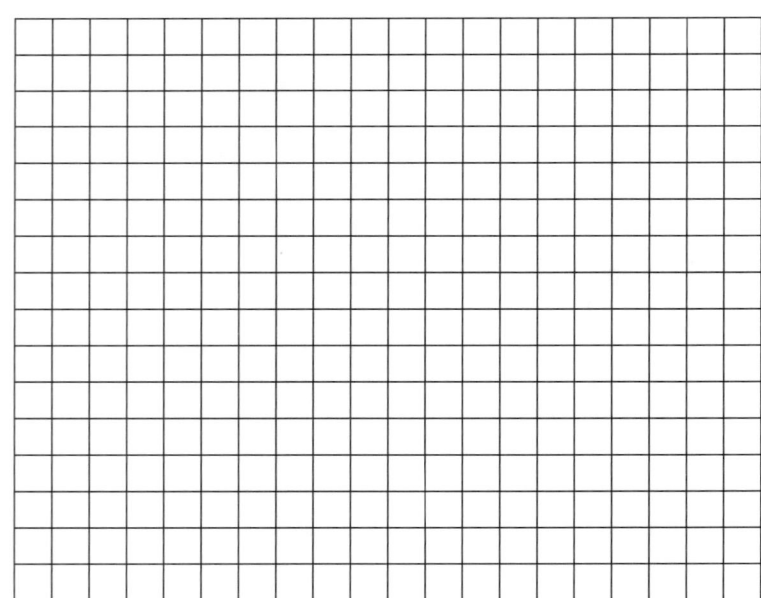

98　　　　　　　　　　　　　　　　　　　National Science Teachers Association

Investigation 6. Energy Transfer by Electric Currents: How Can We Make an Electric Car Move Faster?

This analysis of the data I collected suggests _____

_____

_____

_____

_____

_____

This evidence is important because of several scientific concepts. The first one is _____

_____

_____

_____

_____

_____

 **Review**

Your friends need your help! Review the draft of their investigation reports and give them ideas about how to improve. Use the *peer-review guide* that begins on the next page to guide your review.

Investigation Log

Peer-Review Guide

Section 1: The Investigation	Reviewer Rating		
1. Did the author do a good job of explaining what the investigation was about?	☐ No	☐ Almost	☐ Yes
2. Did the author do a good job of making the **guiding question** clear?	☐ No	☐ Almost	☐ Yes
3. Did the author do a good job of describing what he or she did to **collect data**?	☐ No	☐ Almost	☐ Yes
4. Did the author do a good job describing **how** he or she **analyzed** the data?	☐ No	☐ Almost	☐ Yes
Reviewers: If your group gave the author any "No" or "Almost" ratings, please give the author some advice about what to do to improve this part of his or her investigation report.			

Section 2: The Argument	Reviewer Rating		
1. Does the author's claim provide a clear and detailed **answer** to the guiding question?	☐ No	☐ Almost	☐ Yes
2. Did the author support his or her claim with **scientific evidence?** Scientific evidence includes analyzed data and an explanation of the analysis.	☐ No	☐ Almost	☐ Yes
3. Does the **evidence** that the author uses in his or her argument **support the claim?**	☐ No	☐ Almost	☐ Yes
4. Did the author include enough **evidence** in his or her argument?	☐ No	☐ Almost	☐ Yes
5. Did the author do a good job of **explaining why the evidence** is important (why it matters)?	☐ No	☐ Almost	☐ Yes
6. Is the content of the argument **correct** based on the science concepts we talked about in class?	☐ No	☐ Almost	☐ Yes
Reviewers: If your group gave the author any "No" or "Almost" ratings, please give the author some advice about what to do to improve this part of his or her investigation report.			

Continued

Investigation 6. Energy Transfer by Electric Currents:
How Can We Make an Electric Car Move Faster?

Section 3: Mechanics	Reviewer Rating		
1. **Grammar:** Are the sentences complete? Is there proper subject-verb agreement in each sentence? Are there no run-on sentences?	☐ No	☐ Almost	☐ Yes
2. **Conventions:** Did the author use proper spelling, punctuation, and capitalization?	☐ No	☐ Almost	☐ Yes
3. **Word Choice:** Did the author use the right words in each sentence (for example, *there* vs. *their, to* vs. *too, then* vs. *than*)?	☐ No	☐ Almost	☐ Yes

Reviewers: If your group gave the author any "No" or "Almost" ratings, please give the author some advice about what to do to improve the writing mechanics of his or her investigation report.

General Reviewer Comments

We liked …

We wonder …

Investigation Log

Write Your Final Report

Once you have received feedback from your friends about your draft report, create your final investigation report in the space that follows.

Introduction

_____
_____
_____
_____
_____
_____
_____
_____

Method

_____
_____
_____
_____
_____
_____
_____
_____
_____
_____
_____
_____

Investigation 6. Energy Transfer by Electric Currents: How Can We Make an Electric Car Move Faster?

Argument

Investigation Log

Investigation Report Grading Rubric

Section 1: The Investigation	Missing	Somewhat	Yes
1. The author explained what the investigation was about.	0	1	2
2. The author made the **guiding question** clear.	0	1	2
3. The author **described** what he or she did to **collect data.**	0	1	2
4. The author described **how** he or she **analyzed** the data.	0	1	2

Section 2: The Argument	Missing	Somewhat	Yes
1. The claim includes a clear and detailed **answer** to the guiding question.	0	1	2
2. The author used **scientific evidence** to support the claim. Scientific evidence includes analyzed data and an explanation of the analysis.	0	1	2
3. The evidence **supports the claim.**	0	1	2
4. The author included enough **evidence** in his or her argument.	0	1	2
5. The author **explained why the evidence** is important.	0	1	2
6. The content of the argument is **correct.**	0	1	2

Section 3: Mechanics	Missing	Somewhat	Yes
1. *Grammar:* The sentences are complete. There is proper subject-verb agreement in each sentence. There are no run-on sentences.	0	1	2
2. *Conventions:* The author used proper spelling, punctuation, and capitalization.	0	1	2
3. *Word Choice:* The author used the right words in each sentence (e.g., *there* vs. *their, to* vs. *too, then* vs. *than*).	0	1	2

Teacher Comments

Here are some things I really liked about your report …	Here are some things I think you could do next time to make your report even better …

Total: _____ /26

Checkout Questions

Investigation 6. Energy Transfer by Electric Currents

1. The pictures below show three different propeller cars.

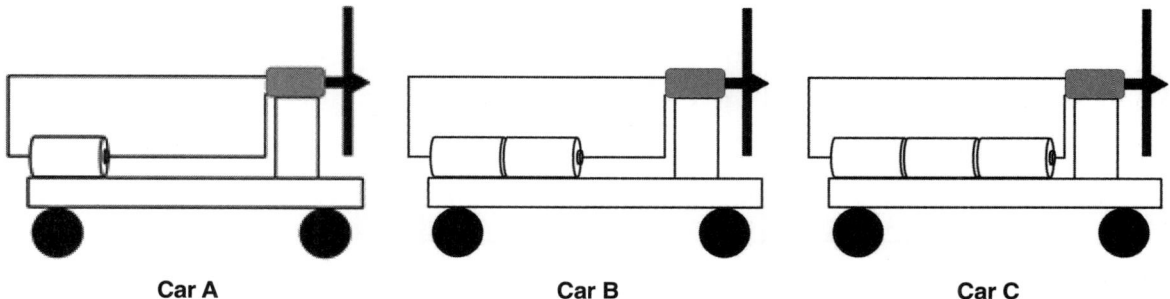

Car A　　　　　　　　Car B　　　　　　　　Car C

Which car would likely move the fastest?

   a. Car A

   b. Car B

   c. Car C

2. The pictures below show three different propeller cars.

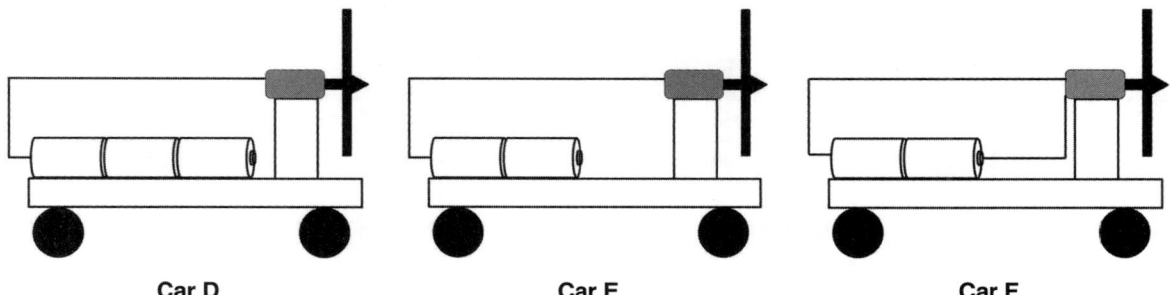

Car D　　　　　　　　Car E　　　　　　　　Car F

Which car would likely move the fastest?

   a. Car D

   b. Car E

   c. Car F

3. Explain your thinking. How does what you know about the *transfer of energy* by electric currents help you make these predictions?

_____

_____

_____

# Checkout Questions

_____
_____
_____
_____
_____
_____
_____
_____
_____
_____
_____
_____
_____

Teacher Scoring Rubric for the Checkout Questions

Level	Description
3	The student can apply the core idea correctly in all cases and can fully explain the transfer of energy.
2	The student can apply the core idea correctly in all cases but cannot fully explain the transfer of energy.
1	The student cannot apply the core idea correctly in all cases but can fully explain the transfer of energy.
0	The student cannot apply the core idea correctly in all cases and cannot explain the transfer of energy.

Section 3
Waves and Their Application in Technologies for Information Transfer

Investigation Log

Investigation 7

Production of Waves: How Does Changing the Mass of an Object That Is Dropped Into Water Affect the Height of the Resulting Wave and the Position of a Floating Toy Boat?

Introduction

Waves are everywhere. One example of a wave is what we see when we drop an object into water. Take a few minutes to explore what happens when you drop a marble into a large round plastic container that is filled with water. As you explore what happens after the marble hits the water, keep track of what you observe and what you are wondering about in the boxes below.

Things I OBSERVED …	Things I WONDER about …

A wave, such as the water wave you just observed, is just another name for a disturbance that travels through a substance or material from one place to another. In this case, you created a disturbance in water, which is a substance, by dropping a marble into it. The disturbance that you created in the water

Investigation Log

with the marble then moved through the rest of the water in all directions. All water waves have several characteristics in common. Scientists can use these characteristics to describe the properties of different water waves that they see.

A water wave always has a high point and a low point. The high point of a wave is called the *crest*, and the low point of a wave is called the *trough*. The distance between the surface of the water and the top of the crest or the distance between the surface of the water and the bottom of the trough is called the *amplitude* of a wave. When people describe one wave as being higher or taller than another wave, they are really talking about the difference in the amplitudes of two different waves.

Water waves also have several crests and troughs. The crests and troughs always follow an alternating pattern. A trough always follows a crest, and a crest always follows a trough. Scientists call the distance between two crests or two troughs the *wavelength* of a wave. Some waves have a short wavelength, and some waves have a long wavelength. Scientists can measure the amplitude and wavelength of a wave and then share this information with other people as a way to describe a specific wave. For example, a scientist might describe the water waves you observed earlier by saying, "That wave had an amplitude of 2 centimeters and a wavelength of 1 centimeter."

Your goal in this investigation is to figure out how the mass of a falling object that creates a disturbance in water affects the amplitude of the resulting wave and the position of a toy boat that is floating in the water. You can create water waves in a clear plastic container filled with water. The object that you will use to create a disturbance in the water is a plastic bag filled with sand. You can change the mass of the plastic bag by adding more sand. You can measure the amplitude, or height, of a water wave and how the toy boat changes position with two different rulers. You may need to tape the rulers to the side of the plastic container to keep them from moving. You may also need to video record the wave as it moves by the ruler so you can measure its amplitude.

As you design and carry out your investigation, be sure to keep in mind that a falling object has energy because it is moving, energy can be transferred between objects when there is a collision, and waves have regular patterns of motion that we can use to describe them.

Things we KNOW from what we read …	What we will NEED to figure out …

Investigation 7. Production of Waves:
How Does Changing the Mass of an Object That Is Dropped Into Water Affect the Height of the Resulting Wave and the Position of a Floating Toy Boat?

Your Task

Use what you know about patterns, cause and effect, and wave properties to design and carry out an investigation to learn more about how dropping objects with different masses into water changes the amplitude of a water wave and the position of a floating toy boat.

The *guiding question* of this investigation is, **How does changing the mass of an object that is dropped into water affect the height of the resulting wave and the position of a floating toy boat?**

Materials

You may use any of the following materials during your investigation:

- Safety glasses or goggles (required)
- Plastic bag
- Sand
- Scale
- 2 rulers
- Toy boat
- Long narrow plastic container filled with water
- Video camera (if available)

Safety Rules

Follow all normal safety rules. In addition, be sure to follow these rules:

- Wear sanitized safety glasses or goggles during setup, investigation activity, and cleanup.
- To avoid a slip or fall hazard, immediately wipe up spills and pick up any items dropped on the floor.
- Wash your hands with soap and water when you are done collecting the data.

Plan Your Investigation

Prepare a plan for your investigation by filling out the chart that follows; this plan is called an *investigation proposal*. Before you start developing your plan, be sure to discuss the following questions with the other members of your group:

- What types of *patterns* might we look for to help answer the guiding question?
- What information do we need to find a relationship between a *cause and an effect*?

Investigation Log

Our guiding question:

This is a picture of how we will set up the equipment:

We will collect the following data:

These are the steps we will follow to collect data:

I approve of this investigation proposal.

Teacher's signature   Date

National Science Teachers Association

Investigation 7. Production of Waves:
How Does Changing the Mass of an Object That Is Dropped Into Water Affect the Height of the Resulting Wave and the Position of a Floating Toy Boat?

Collect Your Data

Keep a record of what you measure or observe during your investigation in the space below.

Analyze Your Data

You will need to analyze the data you collected before you can develop an answer to the guiding question. To do this, create two graphs. Each graph should show the relationship between the cause and the effect.

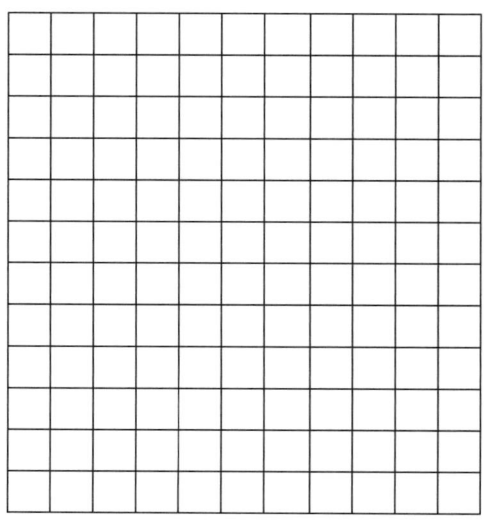 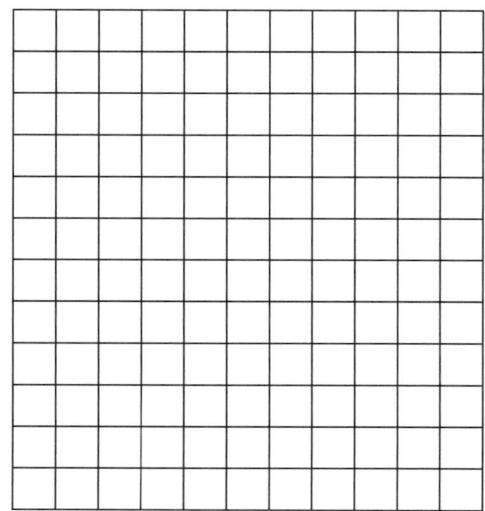

Student Workbook for Argument-Driven Inquiry in **Fourth-Grade Science:** Three-Dimensional Investigations

Investigation Log

Draft Argument

Develop an argument on a whiteboard. It should include the following:

1. A *claim*: Your answer to the guiding question.
2. *Evidence:* An analysis of the data and an explanation of what the analysis means.
3. A *justification of the evidence:* Why your group thinks the evidence is important.

The Guiding Question:	
Our Claim:	
Our Evidence:	Our Justification of the Evidence:

Argumentation Session

Share your argument with your classmates. Be sure to ask them how to make your draft argument better. Keep track of their suggestions in the space below.

Ways to IMPROVE our argument …

Investigation 7. Production of Waves:
How Does Changing the Mass of an Object That Is Dropped Into Water Affect the Height of the Resulting Wave and the Position of a Floating Toy Boat?

Draft Report

Prepare an *investigation report* to share what you have learned. Use the information in this handout and your group's final argument to write a *draft* of your investigation report.

Introduction

We have been studying _____ in class.

Before we started this investigation, we explored _____

_____

_____

We noticed _____

_____

_____

My goal for this investigation was to figure out _____

_____

_____

The guiding question was _____

_____

_____

Method

To gather the data I needed to answer this question, I _____

_____

_____

Student Workbook for Argument-Driven Inquiry in **Fourth-Grade Science**: Three-Dimensional Investigations

Investigation Log

I then analyzed the data I collected by _____

_____

_____

_____

_____

_____

_____

_____

Argument

My claim is _____

_____

_____

_____

_____

The graphs below show _____

_____

_____

_____

National Science Teachers Association

Investigation 7. Production of Waves:
How Does Changing the Mass of an Object That Is Dropped Into Water Affect the Height of the Resulting Wave and the Position of a Floating Toy Boat?

These analyses of the data I collected suggest _____

_____

_____

_____

_____

_____

_____

This evidence is important because of several scientific concepts. The first one is _____

_____

_____

_____

_____

_____

_____

_____

_____

_____

 **Review**

Your friends need your help! Review the draft of their investigation reports and give them ideas about how to improve. Use the *peer-review guide* that begins on the next page to guide your review.

Investigation Log

Peer-Review Guide

Section 1: The Investigation	Reviewer Rating		
1. Did the author do a good job of explaining what the investigation was about?	☐ No	☐ Almost	☐ Yes
2. Did the author do a good job of making the **guiding question** clear?	☐ No	☐ Almost	☐ Yes
3. Did the author do a good job of describing what he or she did to **collect data**?	☐ No	☐ Almost	☐ Yes
4. Did the author do a good job describing **how** he or she **analyzed** the data?	☐ No	☐ Almost	☐ Yes

Reviewers: If your group gave the author any "No" or "Almost" ratings, please give the author some advice about what to do to improve this part of his or her investigation report.

Section 2: The Argument	Reviewer Rating		
1. Does the author's claim provide a clear and detailed **answer** to the guiding question?	☐ No	☐ Almost	☐ Yes
2. Did the author support his or her claim with **scientific evidence?** Scientific evidence includes analyzed data and an explanation of the analysis.	☐ No	☐ Almost	☐ Yes
3. Does the **evidence** that the author uses in his or her argument **support the claim?**	☐ No	☐ Almost	☐ Yes
4. Did the author include enough **evidence** in his or her argument?	☐ No	☐ Almost	☐ Yes
5. Did the author do a good job of **explaining why the evidence** is important (why it matters)?	☐ No	☐ Almost	☐ Yes
6. Is the content of the argument **correct** based on the science concepts we talked about in class?	☐ No	☐ Almost	☐ Yes

Reviewers: If your group gave the author any "No" or "Almost" ratings, please give the author some advice about what to do to improve this part of his or her investigation report.

Continued

Investigation 7. Production of Waves: How Does Changing the Mass of an Object That Is Dropped Into Water Affect the Height of the Resulting Wave and the Position of a Floating Toy Boat?

Section 3: Mechanics	Reviewer Rating		
1. *Grammar:* Are the sentences complete? Is there proper subject-verb agreement in each sentence? Are there no run-on sentences?	☐ No	☐ Almost	☐ Yes
2. *Conventions:* Did the author use proper spelling, punctuation, and capitalization?	☐ No	☐ Almost	☐ Yes
3. *Word Choice:* Did the author use the right words in each sentence (for example, *there* vs. *their, to* vs. *too, then* vs. *than*)?	☐ No	☐ Almost	☐ Yes

Reviewers: If your group gave the author any "No" or "Almost" ratings, please give the author some advice about what to do to improve the writing mechanics of his or her investigation report.

General Reviewer Comments

We liked …

We wonder …

Investigation Log

Write Your Final Report

Once you have received feedback from your friends about your draft report, create your final investigation report in the space that follows.

Introduction

_____
_____
_____
_____
_____
_____
_____

Method

_____
_____
_____
_____
_____
_____
_____
_____
_____
_____

Investigation 7. Production of Waves: How Does Changing the Mass of an Object That Is Dropped Into Water Affect the Height of the Resulting Wave and the Position of a Floating Toy Boat?

Argument

Investigation Log

Investigation Report Grading Rubric

Section 1: The Investigation	Score Missing	Somewhat	Yes
1. The author explained what the investigation was about.	0	1	2
2. The author made the **guiding question** clear.	0	1	2
3. The author **described** what he or she did to **collect data**.	0	1	2
4. The author described **how** he or she **analyzed** the data.	0	1	2

Section 2: The Argument	Score Missing	Somewhat	Yes
1. The claim includes a clear and detailed **answer** to the guiding question.	0	1	2
2. The author used **scientific evidence** to support the claim. Scientific evidence includes analyzed data and an explanation of the analysis.	0	1	2
3. The evidence **supports the claim.**	0	1	2
4. The author included enough **evidence** in his or her argument.	0	1	2
5. The author **explained why the evidence** is important.	0	1	2
6. The content of the argument is **correct.**	0	1	2

Section 3: Mechanics	Score Missing	Somewhat	Yes
1. *Grammar:* The sentences are complete. There is proper subject-verb agreement in each sentence. There are no run-on sentences.	0	1	2
2. *Conventions:* The author used proper spelling, punctuation, and capitalization.	0	1	2
3. *Word Choice:* The author used the right words in each sentence (e.g., *there* vs. *their, to* vs. *too, then* vs. *than*).	0	1	2

Teacher Comments

Here are some things I really liked about your report …	Here are some things I think you could do next time to make your report even better …

Total: _____ /26

Checkout Questions

Investigation 7. Production of Waves

Use the following information to answer questions 1–5. The picture below shows a toy boat floating in a tank of water. There are three balls next to the tank. The balls are the same size, but each one is a different mass.

1. Which ball should you drop into the left side of the tank to make a wave with the *largest* amplitude on the right side of the tank?

    a. The 10 g ball.

    b. The 30 g ball.

    c. The 50 g ball.

    d. The mass of the ball doesn't matter.

2. Which ball should you drop into the left side of the tank to make a wave with the *smallest* amplitude on the right side of the tank?

    a. The 10 g ball.

    b. The 30 g ball.

    c. The 50 g ball.

    d. The mass of the ball doesn't matter.

3. Which ball should you drop into the left side of the tank to make the toy boat move to the right *the most*?

    a. The 10 g ball.

    b. The 30 g ball.

    c. The 50 g ball.

    d. The mass of the ball doesn't matter.

4. Which ball should you drop into the left side of the tank to make the toy boat move to the right *the least*?

    a. The 10 g ball.

    b. The 30 g ball.

    c. The 50 g ball.

    d. The mass of the ball doesn't matter.

Student Workbook for Argument-Driven Inquiry in **Fourth-Grade Science**: Three-Dimensional Investigations

Checkout Questions

5. Explain your thinking. How did you use what you know about *wave properties* and *patterns* to answer these questions?

_____

_____

_____

_____

_____

_____

_____

_____

_____

_____

_____

_____

Teacher Scoring Rubric for the Checkout Questions

Level	Description
3	The student can apply the core idea correctly in all cases and can fully explain the pattern.
2	The student can apply the core idea correctly in all cases but cannot fully explain the pattern.
1	The student cannot apply the core idea correctly in all cases but can fully explain the pattern.
0	The student cannot apply the core idea correctly in all cases and cannot explain the pattern.

Investigation Log

Investigation 8

Characteristics of Sound Waves: How Can We Use the Properties of Waves to Explain Different Sounds?

Introduction

When you hit a solid object, it will often bend a little bit and then move back and forth. This back-and-forth motion is called a *vibration*. An object that is vibrating can produce a sound. Take a few minutes to explore what happens when you hit the end of a tuning fork on a table. As you explore what happens over time, keep track of what you observe and what you are wondering about in the boxes below.

Things I OBSERVED ...	Things I WONDER about ...

Sound is a wave. Sound waves, like all waves, are created when a disturbance travels through a substance or material from one place to another. In this case, you created a sound wave by making a disturbance in the air. You created a disturbance in the air by striking a tuning fork and making the tines vibrate. As the tines moved back and forth, they disturbed the surrounding air. The disturbance in the air then moved out away from the tuning fork in all directions. You are able to hear sound because your ears are able to detect a disturbance as it moves through air.

Investigation Log

We can use a tool called an *oscilloscope* to see sound waves. An *oscilloscope* measures changes in air pressure and then uses this information to create a model of a sound wave. A sound wave has some specific properties when we see it on the screen of an oscilloscope. The model of the sound wave will always have a high point and a low point. The high point of the wave is called the crest, and the low point of the wave is called the *trough*. The point between the high point and the low point of the wave is called the *resting point*. The line you see on the oscilloscope when there is no sound runs through the resting point. The distance between the resting point and the top of the crest or the distance between the resting point and bottom of the trough is called the *amplitude* of a wave. The distance between two crests separated by a trough or two troughs separated by a crest is called the *wavelength* of a wave. We can use these properties of waves to describe any sound wave that we see on an oscilloscope.

Your goal in this investigation is to figure out how the characteristics of the sounds that we are able to hear are related to the properties of a sound wave that we see on an oscilloscope. You will focus on two characteristics of a sound during this investigation: (1) pitch and (2) volume. You can make sounds at different pitches by using different tuning forks. Some tuning forks produce a high-pitched sound, and some produce a low-pitched sound. You can increase the volume of a sound by striking a tuning force with different amounts of force. When you strike a tuning fork with a small force you will produce a soft sound, and when you strike a tuning fork with a large force you will produce a loud sound. You can also make a sound louder or softer by moving the tuning fork closer to or farther away from the oscilloscope microphone. Your teacher will show you how to use an oscilloscope before you begin designing your investigation.

As you design and carry out your investigation, be sure to keep in mind that sound waves are a disturbance in the air that travels from one place to another, vibrating objects create disturbances in the air, and waves have specific properties that we can use to describe them.

Things we KNOW from what we read …	What we will NEED to figure out …

Investigation 8. Characteristics of Sound Waves:
How Can We Use the Properties of Waves to Explain Different Sounds?

Your Task

Use what you know about waves, patterns, and measurement scales to design and carry out an investigation to learn more about how the pitch and volume of a sound are related to the amplitude and wavelength of a sound wave that is seen on an oscilloscope.

The *guiding question* of this investigation is, ***How can we use the properties of waves to explain different sounds?***

Materials

You may use any of the following materials during your investigation:

- Safety glasses or goggles (required)
- Set of 8 tuning forks
- Microphone
- Ruler
- Computer

In addition, you will use an online version of an oscilloscope, called Virtual Oscilloscope, in your investigation; it can be accessed at *https://academo.org/demos/virtual-oscilloscope*.

Safety Rules

Follow all normal safety rules. In addition, be sure to follow these rules:

- Wear sanitized safety glasses or goggles during setup, investigation activity, and cleanup.
- Do not hit other people with the tuning forks.
- Wash your hands with soap and water when you are done collecting the data.

Plan Your Investigation

Prepare a plan for your investigation by filling out the chart that follows; this plan is called an *investigation proposal*. Before you start developing your plan, be sure to discuss the following questions with the other members of your group:

- What types of *patterns* might we look for to help answer the guiding question?
- How can we find patterns by taking **measurements**, and what **scale** should we use?

Investigation Log

Our guiding question:

We will collect the following data:

These are the steps we will follow to collect data:

I approve of this investigation proposal.

_____ _____
Teacher's signature   Date

Investigation 8. Characteristics of Sound Waves:
How Can We Use the Properties of Waves to Explain Different Sounds?

Collect Your Data

Keep a record of what you measure or observe during your investigation in the space below.

Analyze Your Data

You will need to analyze the data you collected before you can develop an answer to the guiding question. To do this, create two graphs. Each graph should show a relationship between something you changed and what you measured or observed as a result of what you changed.

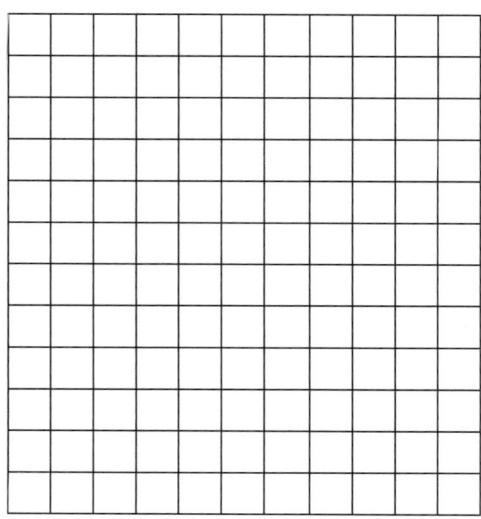
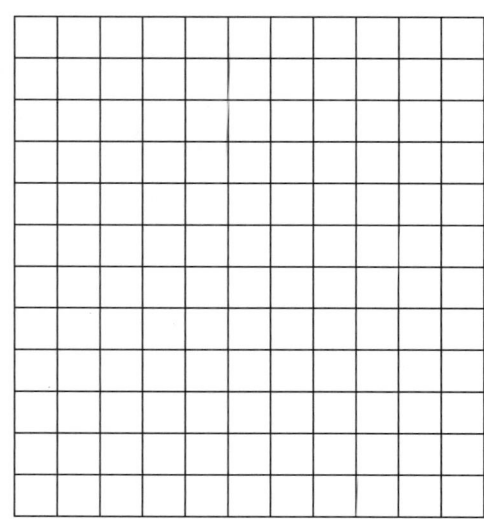

Investigation Log

Draft Argument

Develop an argument on a whiteboard. It should include the following:

1. A *claim*: Your answer to the guiding question.
2. *Evidence*: An analysis of the data and an explanation of what the analysis means.
3. A *justification of the evidence*: Why your group thinks the evidence is important.

The Guiding Question:	
Our Claim:	
Our Evidence:	Our Justification of the Evidence:

Argumentation Session

Share your argument with your classmates. Be sure to ask them how to make your draft argument better. Keep track of their suggestions in the space below.

Ways to IMPROVE our argument …

Investigation 8. Characteristics of Sound Waves:
How Can We Use the Properties of Waves to Explain Different Sounds?

Draft Report

Prepare an investigation report to share what you have learned. Use the information in this handout and your group's final argument to write a draft of your investigation report.

Introduction

We have been studying _____ in class.

Before we started this investigation, we explored _____

_____

_____

We noticed _____

_____

_____

_____

_____

My goal for this investigation was to figure out _____

_____

_____

_____

_____

_____

The guiding question was _____

_____

_____

_____

_____

_____

Investigation Log

Method

To gather the data I needed to answer this question, I _____

_____

_____

_____

_____

_____

_____

_____

I then analyzed the data I collected by _____

_____

_____

_____

_____

_____

Argument

My claim is _____

_____

_____

_____

The graphs that follow show _____

_____

_____

_____

_____

_____

Investigation 8. Characteristics of Sound Waves:
How Can We Use the Properties of Waves to Explain Different Sounds?

_____
_____
_____

These analyses of the data I collected suggest _____
_____
_____
_____
_____
_____
_____

This evidence is important because of several scientific concepts. The first one is _____
_____
_____
_____
_____
_____
_____

Review

Your friends need your help! Review the draft of their investigation reports and give them ideas about how to improve. Use the *peer-review guide* that begins on the next page to guide your review.

Investigation 8. Characteristics of Sound Waves:
How Can We Use the Properties of Waves to Explain Different Sounds?

Peer-Review Guide

Section 1: The Investigation		Reviewer Rating	
1. Did the author do a good job of explaining what the investigation was about?	☐ No	☐ Almost	☐ Yes
2. Did the author do a good job of making the **guiding question** clear?	☐ No	☐ Almost	☐ Yes
3. Did the author do a good job of describing what he or she did to **collect data**?	☐ No	☐ Almost	☐ Yes
4. Did the author do a good job describing **how** he or she **analyzed** the data?	☐ No	☐ Almost	☐ Yes
Reviewers: If your group gave the author any "No" or "Almost" ratings, please give the author some advice about what to do to improve this part of his or her investigation report.			

Section 2: The Argument		Reviewer Rating	
1. Does the author's claim provide a clear and detailed **answer** to the guiding question?	☐ No	☐ Almost	☐ Yes
2. Did the author support his or her claim with **scientific evidence?** Scientific evidence includes analyzed data and an explanation of the analysis.	☐ No	☐ Almost	☐ Yes
3. Does the **evidence** that the author uses in his or her argument **support the claim?**	☐ No	☐ Almost	☐ Yes
4. Did the author include enough **evidence** in his or her argument?	☐ No	☐ Almost	☐ Yes
5. Did the author do a good job of **explaining why the evidence** is important (why it matters)?	☐ No	☐ Almost	☐ Yes
6. Is the content of the argument **correct** based on the science concepts we talked about in class?	☐ No	☐ Almost	☐ Yes
Reviewers: If your group gave the author any "No" or "Almost" ratings, please give the author some advice about what to do to improve this part of his or her investigation report.			

Continued

Investigation Log

Section 3: Mechanics	Reviewer Rating		
1. **Grammar:** Are the sentences complete? Is there proper subject-verb agreement in each sentence? Are there no run-on sentences?	☐ No	☐ Almost	☐ Yes
2. **Conventions:** Did the author use proper spelling, punctuation, and capitalization?	☐ No	☐ Almost	☐ Yes
3. **Word Choice:** Did the author use the right words in each sentence (for example, *there* vs. *their, to* vs. *too, then* vs. *than*)?	☐ No	☐ Almost	☐ Yes

Reviewers: If your group gave the author any "No" or "Almost" ratings, please give the author some advice about what to do to improve the writing mechanics of his or her investigation report.

General Reviewer Comments

We liked …

We wonder …

Investigation 8. Characteristics of Sound Waves:
How Can We Use the Properties of Waves to Explain Different Sounds?

Write Your Final Report

Once you have received feedback from your friends about your draft report, create your final investigation report in the space that follows.

Introduction

_____
_____
_____
_____
_____
_____
_____

Method

_____
_____
_____
_____
_____
_____
_____
_____
_____
_____
_____

Investigation Log

Argument

Investigation 8. Characteristics of Sound Waves:
How Can We Use the Properties of Waves to Explain Different Sounds?

Investigation Report Grading Rubric

Section 1: The Investigation	Score Missing	Somewhat	Yes
1. The author explained what the investigation was about.	0	1	2
2. The author made the **guiding question** clear.	0	1	2
3. The author **described** what he or she did to **collect data.**	0	1	2
4. The author described **how** he or she **analyzed** the data.	0	1	2
Section 2: The Argument	**Score Missing**	**Somewhat**	**Yes**
1. The claim includes a clear and detailed **answer** to the guiding question.	0	1	2
2. The author used **scientific evidence** to support the claim. Scientific evidence includes analyzed data and an explanation of the analysis.	0	1	2
3. The evidence **supports the claim.**	0	1	2
4. The author included enough **evidence** in his or her argument.	0	1	2
5. The author **explained why the evidence** is important.	0	1	2
6. The content of the argument is **correct.**	0	1	2
Section 3: Mechanics	**Score Missing**	**Somewhat**	**Yes**
1. *Grammar:* The sentences are complete. There is proper subject-verb agreement in each sentence. There are no run-on sentences.	0	1	2
2. *Conventions:* The author used proper spelling, punctuation, and capitalization.	0	1	2
3. *Word Choice:* The author used the right words in each sentence (e.g., *there* vs. *their, to* vs. *too, then* vs. *than*).	0	1	2

Teacher Comments

Here are some things I really liked about your report …	Here are some things I think you could do next time to make your report even better …

Total: _____ /26

Student Workbook for Argument-Driven Inquiry in **Fourth-Grade Science:** Three-Dimensional Investigations

Checkout Questions

Investigation 8. Characteristics of Sound Waves

1. The image below shows a sound wave when viewed using an oscilloscope. Draw how the properties of this sound wave would change if someone made the sound *louder*.

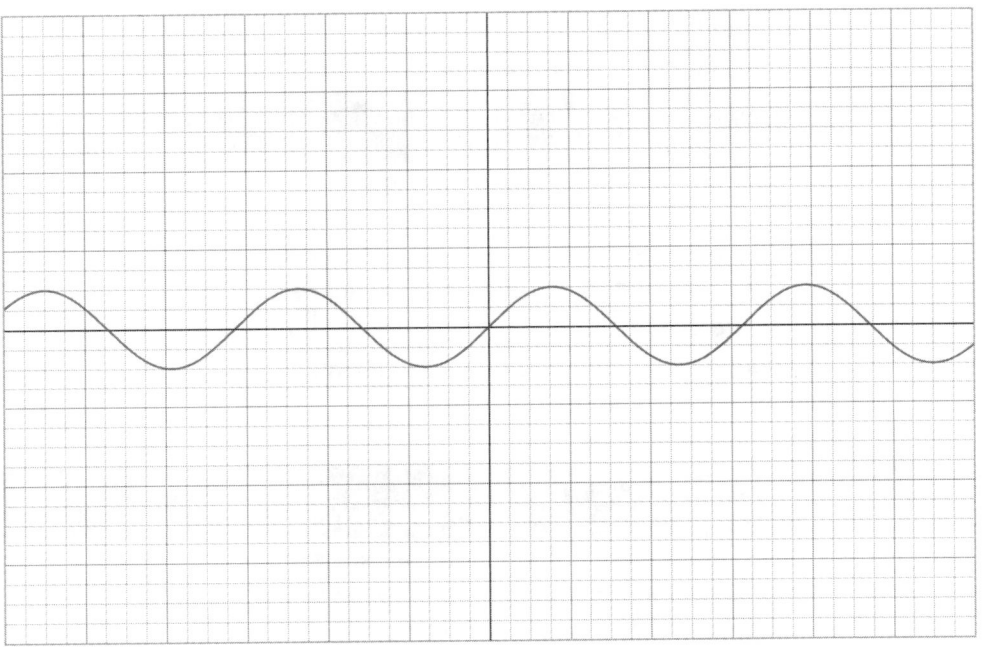

2. The image below shows another sound wave when viewed using an oscilloscope. Draw how the properties of this sound wave would change if someone made the pitch *higher*.

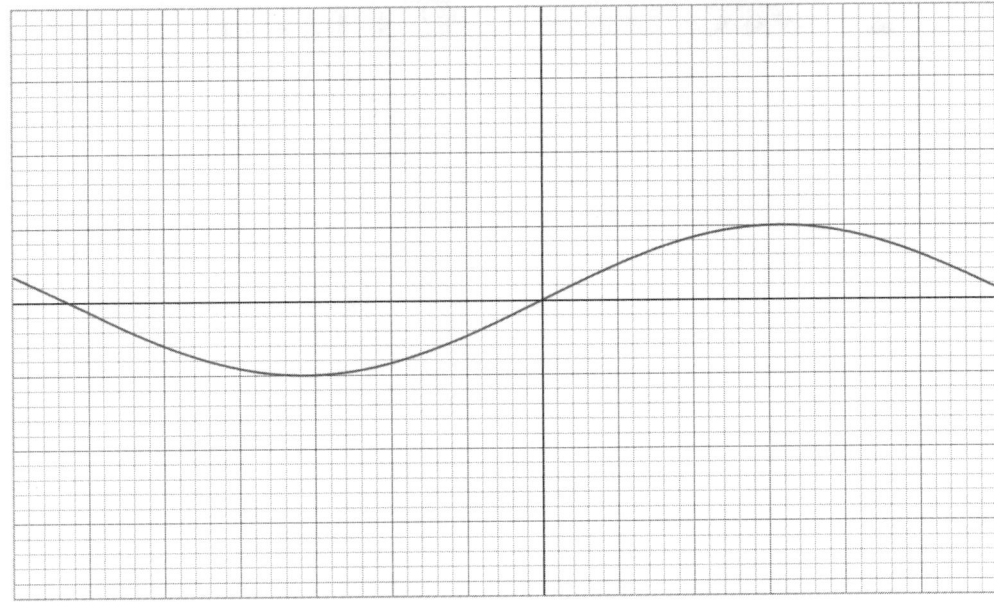

Investigation 8. Characteristics of Sound Waves:
How Can We Use the Properties of Waves to Explain Different Sounds?

3. The image below shows another sound wave when viewed using an oscilloscope. Draw how the properties of this sound wave would change if someone made the pitch *lower*.

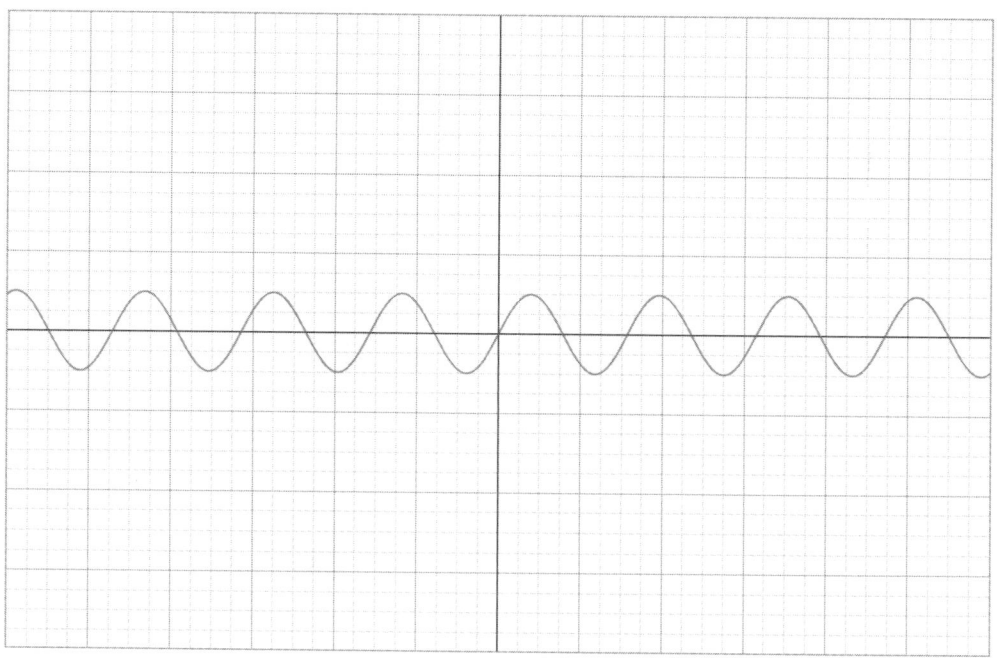

4. Explain your thinking. How did you use what you know about *wave properties* and *patterns* to answer these questions?

_____
_____
_____
_____
_____
_____
_____

Teacher Scoring Rubric for the Checkout Questions

Level	Description
3	The student can apply the core idea correctly in all cases and can fully explain the pattern.
2	The student can apply the core idea correctly in all cases but cannot fully explain the pattern.
1	The student cannot apply the core idea correctly in all cases but can fully explain the pattern.
0	The student cannot apply the core idea correctly in all cases and cannot explain the pattern.

Investigation Log

Investigation 9

Light and Reflection: What Is the Relationship Between the Location of an Object and the Location of Its Image in a Mirror?

Introduction

Mirrors can be found in our homes and in the places we shop, and sometimes they are even mounted on a pole so people can use them to see around a corner. Take a few minutes to see what you have to do to be able to use a mirror to see behind a folder. Be sure to keep track of what you observe and what you are wondering about as you use the mirror in the boxes below.

Things I OBSERVED …	Things I WONDER about …

We can only see an object that we are looking at when light from that object travels to our eyes. Once light reaches our eyes, signals are sent to our brain, and our brain interprets the light that hits our eyes in a way that allows us to see the appearance, location, and movement of the object that we are looking at. Therefore, for you to be able to see an object, the object must either produce its own light or reflect light that was produced by something else. You are able to see your friends because light is hitting them, bouncing off them, and then traveling to your eye.

Investigation Log

Pick a person in your class. Now take a moment to look at that person. To see that person, you must move your head so your eyes are pointed toward that person. The direction you are looking is called the *line of sight*. If you wish to view the top of a person's head, then you must move your head so there is a line of sight between your eyes and the top of that person's head. If you wish to see your classmate's feet, then you must move your head so there is a line of sight between your eyes and his or her feet. And if you wish to see an image of a friend in a mirror, then you must move your head until there is a line of sight between your eyes and the location of your friend's image. To be able to see any object, you must have a line of sight between your eyes and that object. When you have a line of sight, light is able to travel from that object to your eye; if you do not have a line of sight, light is unable to travel from the object to your eye.

Your goal in this investigation is to figure out how to predict where an image of a pencil in a mirror will be given the location of the pencil and your line of sight. To determine the location of the pencil image, you will need to use two pencils that are inserted into two different rubber stoppers (so the pencils can stand upright on their own). Place one of these pencils (we will call it pencil A) in front of a mirror. Look at the image of pencil A in the mirror. Place the second pencil (pencil B) behind the mirror so the portion of pencil B that you can see is aligned with the image of pencil A. From there, change your eye position and see if the image of pencil A is aligned with the location of pencil B. If it is not, move pencil B so it is aligned with the image of pencil A that you see. Once you are able to change your eye position without having to move pencil B to align it with the image of pencil A, you have found the image location of pencil A. You can then record where the image is located in relationship to the mirror. From there, you can change the location of pencil A and repeat this process. Your teacher will show you how to do this before you begin designing your investigation. As you carry out your investigation, be sure to keep in mind that we can only see objects that produce or reflect light and we can only see an object when there is a direct line of sight between our eyes and that object.

Things we KNOW from what we read …	What we will NEED to figure out …

144 National Science Teachers Association

Investigation 9. Light and Reflection:
What Is the Relationship Between the Location of an Object and the Location of Its Image in a Mirror?

Your Task

Use what you know about light, cause and effect, and patterns to design and carry out an investigation to figure out how to predict where the image of an object will be located in a mirror.

The *guiding question* of this investigation is, ***What is the relationship between the location of an object and the location of its image in a mirror?***

Materials

You may use any of the following materials during your investigation:

- Safety glasses or goggles (required)
- Plane mirror
- 2 pencils (each inserted into a different rubber stopper)
- 1-centimeter grid paper or coordinate paper
- 2 mirror stands
- Protractor
- Ruler

Safety Rules

Follow all normal safety rules. In addition, be sure to follow these rules:

- Wear sanitized safety glasses or goggles during setup, investigation activity, and cleanup.
- Wash your hands with soap and water when you are done collecting the data.

Plan Your Investigation

Prepare a plan for your investigation by filling out the chart that follows; this plan is called an *investigation proposal*. Before you start developing your plan, be sure to discuss the following questions with the other members of your group:

- What types of **patterns** might we look for to help answer the guiding question?
- What information do we need to find a relationship between a **cause** and an **effect**?

Investigation Log

Our guiding question:

This is a picture of how we will set up the equipment:

We will collect the following data:

These are the steps we will follow to collect data:

I approve of this investigation proposal.

_____ _____
Teacher's signature   Date

Investigation 9. Light and Reflection:
What Is the Relationship Between the Location of an Object and the Location of Its Image in a Mirror?

Collect Your Data

Keep a record of what you measure or observe during your investigation in the space below.

Analyze Your Data

You will need to analyze the data you collected before you can develop an answer to the guiding question. To do this, create a table or graph that shows the relationship between what you changed and what you measured or observed as a result of what you changed.

Investigation Log

Draft Argument

Develop an argument on a whiteboard. It should include the following:

1. A *claim:* Your answer to the guiding question.
2. *Evidence:* An analysis of the data and an explanation of what the analysis means.
3. A *justification of the evidence:* Why your group thinks the evidence is important.

The Guiding Question:	
Our Claim:	
Our Evidence:	Our Justification of the Evidence:

Argumentation Session

Share your argument with your classmates. Be sure to ask them how to make your draft argument better. Keep track of their suggestions in the space below.

Ways to IMPROVE our argument …

Investigation 9. Light and Reflection:
What Is the Relationship Between the Location of an Object and the Location of Its Image in a Mirror?

Draft Report

Prepare an *investigation report* to share what you have learned. Use the information in this handout and your group's final argument to write a *draft* of your investigation report.

Introduction

We have been studying _____ in class. Before we

started this investigation, we explored _____

_____

_____

We noticed _____

_____

_____

My goal for this investigation was to figure out _____

_____

_____

The guiding question was _____

_____

_____

Method

To gather the data I needed to answer this question, I _____

_____

_____

_____

_____

_____

Investigation Log

I then analyzed the data I collected by _____
_____
_____

Argument

My claim is _____
_____
_____

The _____ below includes information about
_____
_____

This analysis of the data I collected suggests _____
_____
_____
_____
_____

Investigation 9. Light and Reflection:
What Is the Relationship Between the Location of an Object and the Location of Its Image in a Mirror?

This evidence is based on several important scientific concepts. The first one is _____

_____

_____

_____

_____

 **Review**

Your friends need your help! Review the draft of their investigation reports and give them ideas about how to improve. Use the *peer-review guide* that begins on the next page to guide your review.

Investigation Log

Peer-Review Guide

Section 1: The Investigation		Reviewer Rating	
1. Did the author do a good job of explaining what the investigation was about?	☐ No	☐ Almost	☐ Yes
2. Did the author do a good job of making the **guiding question** clear?	☐ No	☐ Almost	☐ Yes
3. Did the author do a good job of describing what he or she did to **collect data**?	☐ No	☐ Almost	☐ Yes
4. Did the author do a good job describing **how** he or she **analyzed** the data?	☐ No	☐ Almost	☐ Yes

Reviewers: If your group gave the author any "No" or "Almost" ratings, please give the author some advice about what to do to improve this part of his or her investigation report.

Section 2: The Argument		Reviewer Rating	
1. Does the author's claim provide a clear and detailed **answer** to the guiding question?	☐ No	☐ Almost	☐ Yes
2. Did the author support his or her claim with **scientific evidence**? Scientific evidence includes analyzed data and an explanation of the analysis.	☐ No	☐ Almost	☐ Yes
3. Does the **evidence** that the author uses in his or her argument **support the claim**?	☐ No	☐ Almost	☐ Yes
4. Did the author include enough **evidence** in his or her argument?	☐ No	☐ Almost	☐ Yes
5. Did the author do a good job of **explaining why the evidence** is important (why it matters)?	☐ No	☐ Almost	☐ Yes
6. Is the content of the argument **correct** based on the science concepts we talked about in class?	☐ No	☐ Almost	☐ Yes

Reviewers: If your group gave the author any "No" or "Almost" ratings, please give the author some advice about what to do to improve this part of his or her investigation report.

Continued

Investigation 9. Light and Reflection:
What Is the Relationship Between the Location of an Object and the Location of Its Image in a Mirror?

Section 3: Mechanics		Reviewer Rating	
1. *Grammar:* Are the sentences complete? Is there proper subject-verb agreement in each sentence? Are there no run-on sentences?	☐ No	☐ Almost	☐ Yes
2. *Conventions:* Did the author use proper spelling, punctuation, and capitalization?	☐ No	☐ Almost	☐ Yes
3. *Word Choice:* Did the author use the right words in each sentence (for example, *there* vs. *their, to* vs. *too, then* vs. *than*)?	☐ No	☐ Almost	☐ Yes

Reviewers: If your group gave the author any "No" or "Almost" ratings, please give the author some advice about what to do to improve the writing mechanics of his or her investigation report.

General Reviewer Comments

We liked …

We wonder …

Investigation Log

Write Your Final Report

Once you have received feedback from your friends about your draft report, create your final investigation report in the space that follows.

Introduction

___

Method

___

Investigation 9. Light and Reflection: What Is the Relationship Between the Location of an Object and the Location of Its Image in a Mirror?

Argument

_____
_____
_____
_____
_____
_____
_____
_____
_____

_____
_____
_____
_____
_____
_____

Investigation Log

Investigation Report Grading Rubric

Section 1: The Investigation	Score Missing	Somewhat	Yes
1. The author explained what the investigation was about.	0	1	2
2. The author made the **guiding question** clear.	0	1	2
3. The author **described** what he or she did to **collect data.**	0	1	2
4. The author described **how** he or she **analyzed** the data.	0	1	2

Section 2: The Argument	Score Missing	Somewhat	Yes
1. The claim includes a clear and detailed **answer** to the guiding question.	0	1	2
2. The author used **scientific evidence** to support the claim. Scientific evidence includes analyzed data and an explanation of the analysis.	0	1	2
3. The evidence **supports the claim.**	0	1	2
4. The author included enough **evidence** in his or her argument.	0	1	2
5. The author **explained why the evidence** is important.	0	1	2
6. The content of the argument is **correct.**	0	1	2

Section 3: Mechanics	Score Missing	Somewhat	Yes
1. *Grammar:* The sentences are complete. There is proper subject-verb agreement in each sentence. There are no run-on sentences.	0	1	2
2. *Conventions:* The author used proper spelling, punctuation, and capitalization.	0	1	2
3. *Word Choice:* The author used the right words in each sentence (e.g., *there* vs. *their, to* vs. *too, then* vs. *than*).	0	1	2

Teacher Comments

Here are some things I really liked about your report …	Here are some things I think you could do next time to make your report even better …

Total: _____ /26

Checkout Questions

Investigation 9. Light and Reflection

1. The picture below shows a person standing in a room from above. That person cannot see object A from where she is standing but can see an image of it in the mirror. Create a model that explains how that person is able to see the image of object A.

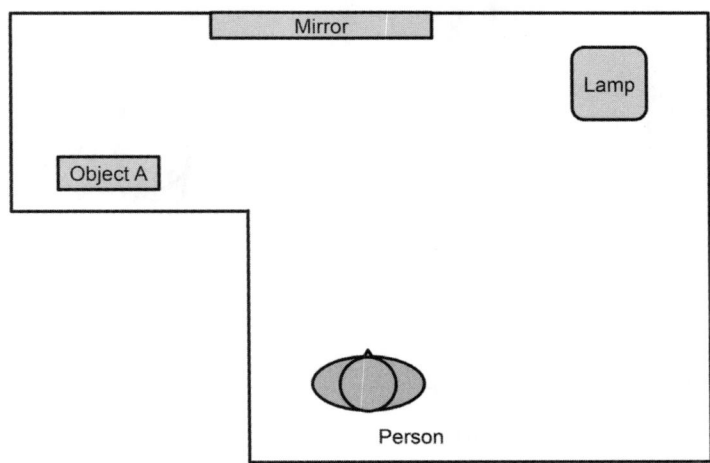

2. Explain your thinking. What cause-and-effect relationship allows people to see images of objects in mirrors?

_____

_____

_____

_____

_____

_____

Teacher Scoring Rubric for Checkout Questions 1 and 2

Level	Description
3	The student can apply the core idea correctly in all cases and can fully explain the cause-and-effect relationship.
2	The student can apply the core idea correctly in all cases but cannot fully explain the cause-and-effect relationship.
1	The student cannot apply the core idea correctly in all cases but can fully explain the cause-and-effect relationship.
0	The student cannot apply the core idea correctly in all cases and cannot explain the cause-and-effect relationship.

Checkout Questions

3. The picture below shows a person standing in front of a mirror from above. That person can see images of object A, object B, and object C in the mirror. Draw where you think the images A, B, and C are located on the picture below.

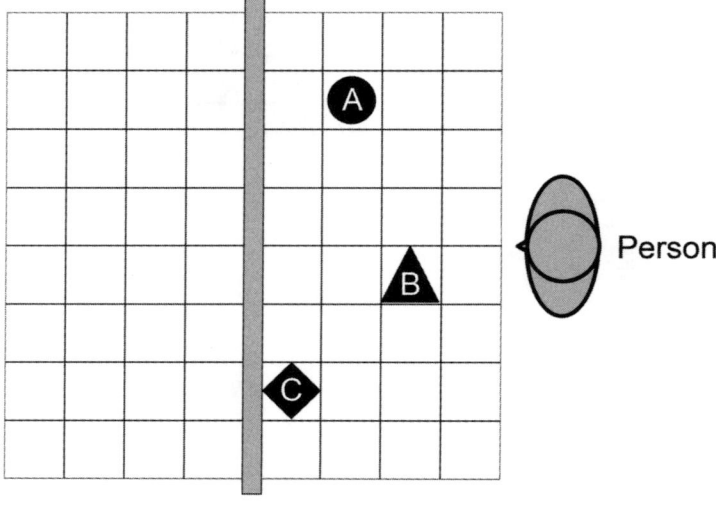

Mirror

4. Explain your thinking. What *pattern* from your investigation allows you to predict the location of an image in a mirror?

_____
_____
_____
_____
_____
_____
_____
_____
_____

Teacher Scoring Rubric for Checkout Questions 3 and 4

Level	Description
3	The student can apply the core idea correctly in all cases and can fully explain the pattern.
2	The student can apply the core idea correctly in all cases but cannot fully explain the pattern.
1	The student cannot apply the core idea correctly in all cases but can fully explain the pattern.
0	The student cannot apply the core idea correctly in all cases and cannot explain the pattern.

Investigation Log

Investigation 10

Patterns and the Transfer of Information: How Can We Use an Electric Circuit to Accurately Transfer Information About the Content of a Picture?

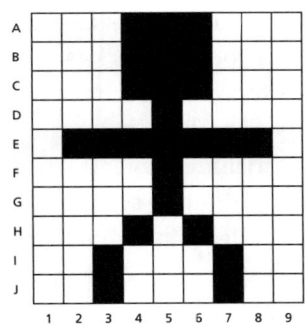

Introduction

We often need to be able to send information over long distances. Engineers have invented phones and computers that can convert our voices or pictures into digital information using a code and then send it to other people. These same devices can also receive digital information, decode it, and convert it back into a voice or an image. For this process to work, the devices we use to send information must use the same code. Take a few minutes to see if you can develop a code that you could use to send a friend the image pictured above. As you create your code, keep track of what it is and what you are wondering about in the boxes below.

Student Workbook for Argument-Driven Inquiry in **Fourth-Grade Science**: Three-Dimensional Investigations

Investigation Log

Computers and cell phones use a code based on a pattern to transfer information. An example of a code that is based on a pattern is called Morse code. This code uses a pattern of dots and dashes to represent each letter in the alphabet. For example, the pattern –••• means B, the pattern •– means A, and the pattern – means T. The word, BAT, is therefore represented as –••••– –. A message that includes several words, as a result, can be sent between two people using long (–) and short (•) signals of light or sound if both people know Morse code. Another example of a code that uses a pattern is called binary code. Binary code is a coding system that uses the digits 0 and 1 to represent a letter, number, or other character in a computer or other electronic device.

An electric current needs a path or loop to travel from place to place. This path or loop is called a *closed circuit*. People can make a closed circuit by linking batteries, switches, and bulbs together with pieces of copper wire or copper tape. If there is a gap anywhere in the circuit or if the electrical components are not connected in the right way, electricity cannot travel through the circuit. This is how switches work. When a switch is in the "on" position, the circuit is closed and electricity can travel through the circuit and the bulb will produce light. When the switch is in the "off" position, however, the circuit is broken and no electricity can flow through the circuit, so the bulb does not produce light. A person can send a message to someone else by simply turning a switch on or off to start or stop the flow of electricity in a circuit if the circuit includes a switch at one location and a bulb at the other location and if the people at both locations use the same code.

Your goal in this investigation is to figure out how to transfer information about the content of a picture that you create from one person to another using an electric circuit. You will have access to copper tape, two different colors of LED lights, and two coin batteries. Your teacher will show you how to make a basic closed circuit on a piece of paper using these materials. You will need to generate and compare multiple solutions for using patterns to transfer information to determine which solution is best. As you design your investigation, keep in mind that batteries store electrical energy, an electric current is movement of electrical energy from one place to another, and patterns can be used to transfer information.

Things we KNOW from what we read …	What we will NEED to figure out …

Investigation 10. Patterns and the Transfer of Information: How Can We Use an Electric Circuit to Accurately Transfer Information About the Content of a Picture?

Your Task

Use what you know about electric currents and patterns to generate and compare multiple solutions for transferring information accurately between people using an electric circuit. The *guiding question* of this investigation is, ***How can we use an electric circuit to accurately transfer information about the content of a picture?***

Materials

You may use any of the following materials during your investigation:

- Safety glasses or goggles (required)
- 2 coin batteries
- 2 LED lights
- Copper tape, 40"
- Circuit pattern paper
- Folder
- Scissors
- Paper with 5 x 5 grids (if needed)

Safety Rules

Follow all normal safety rules. In addition, be sure to follow these rules:

- Wear sanitized safety glasses or goggles during setup, investigation activity, and cleanup.
- Be careful when handling sharp tools and materials, because they can cut or puncture skin.
- Wash your hands with soap and water when you are done collecting the data.

Plan Your Investigation

Prepare a plan for your investigation by filling out the chart that follows; this plan is called an *investigation proposal*. Before you start developing your plan, be sure to discuss the following questions with the other members of your group:

- How can we use **patterns** to transfer information accurately between people?
- What are the parts of a closed-circuit **system**, and how do they interact with each other?

Our guiding question:

This is a picture of how we will set up the equipment:

We will collect the following data:

These are the steps we will follow to collect data:

I approve of this investigation proposal.

_____ _____
Teacher's signature   Date

National Science Teachers Association

Investigation 10. Patterns and the Transfer of Information:
How Can We Use an Electric Circuit to Accurately Transfer Information About the Content of a Picture?

Collect Your Data

Keep a record of each picture you sent and received during your investigation in the space below. Be sure to include information about the number of errors you made.

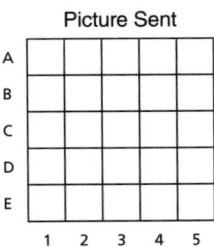 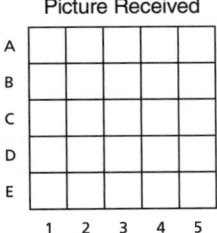

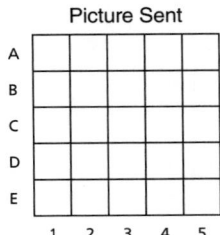 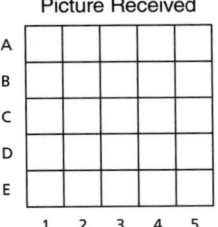

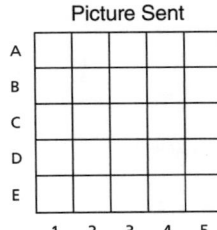 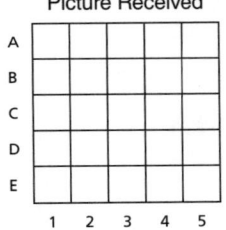

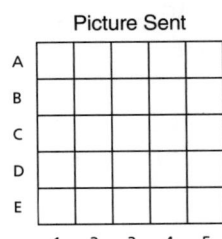 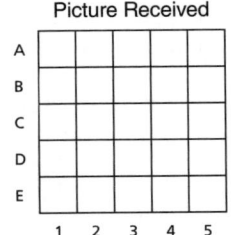

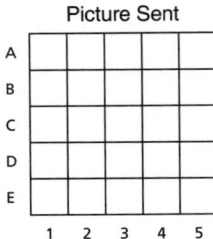 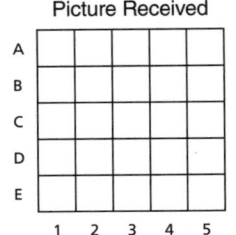

Investigation Log

Analyze Your Data

You will need to analyze the data you collected before you can develop an answer to the guiding question. To do this, create a table or graph that allows you to compare the accuracy of your different solutions. You can also draw pictures.

Draft Argument

Develop an argument on a whiteboard. It should include the following:

1. A *claim:* Your answer to the guiding question.

2. *Evidence:* An analysis of the data and an explanation of what the analysis means.

3. A *justification of the evidence:* Why your group thinks the evidence is important.

The Guiding Question:	
Our Claim:	
Our Evidence:	Our Justification of the Evidence:

Investigation 10. Patterns and the Transfer of Information:
How Can We Use an Electric Circuit to Accurately Transfer Information About the Content of a Picture?

Argumentation Session

Share your argument with your classmates. Be sure to ask them how to make your draft argument better. Keep track of their suggestions in the space below.

Ways to IMPROVE our argument ...

Draft Report

Prepare an *investigation report* to share what you have learned. Use the information in this handout and your group's final argument to write a *draft* of your investigation report.

Introduction

We have been studying _____ in class.

Before we started this investigation, we explored _____

_____

Investigation Log

_____

_____

We noticed _____

_____

_____

My goal for this investigation was to figure out _____

_____

_____

_____

The guiding question was _____

_____

_____

_____

Method

To gather the data I needed to answer this question, I _____

_____

_____

_____

_____

_____

_____

I then analyzed the data I collected by _____

_____

_____

Investigation 10. Patterns and the Transfer of Information: How Can We Use an Electric Circuit to Accurately Transfer Information About the Content of a Picture?

_____
_____
_____
_____
_____

Argument

My claim is _____

_____
_____
_____
_____

The _____ below includes information about

_____
_____
_____

Investigation Log

This analysis of the data I collected suggests _____
_____
_____
_____
_____

This evidence is based on several important scientific concepts. The first one is _____
_____
_____
_____
_____
_____
_____
_____
_____
_____
_____
_____

Review

Your friends need your help! Review the draft of their investigation reports and give them ideas about how to improve. Use the *peer-review guide* that begins on the next page to guide your review.

Investigation 10. Patterns and the Transfer of Information:
How Can We Use an Electric Circuit to Accurately Transfer Information About the Content of a Picture?

Peer-Review Guide

Section 1: The Investigation	Reviewer Rating		
1. Did the author do a good job of explaining what the investigation was about?	☐ No	☐ Almost	☐ Yes
2. Did the author do a good job of making the **guiding question** clear?	☐ No	☐ Almost	☐ Yes
3. Did the author do a good job of describing what he or she did to **collect data**?	☐ No	☐ Almost	☐ Yes
4. Did the author do a good job describing **how** he or she **analyzed** the data?	☐ No	☐ Almost	☐ Yes
Reviewers: If your group gave the author any "No" or "Almost" ratings, please give the author some advice about what to do to improve this part of his or her investigation report.			

Section 2: The Argument	Reviewer Rating		
1. Does the author's claim provide a clear and detailed **answer** to the guiding question?	☐ No	☐ Almost	☐ Yes
2. Did the author support his or her claim with **scientific evidence**? Scientific evidence includes analyzed data and an explanation of the analysis.	☐ No	☐ Almost	☐ Yes
3. Does the **evidence** that the author uses in his or her argument **support the claim**?	☐ No	☐ Almost	☐ Yes
4. Did the author include enough **evidence** in his or her argument?	☐ No	☐ Almost	☐ Yes
5. Did the author do a good job of **explaining why the evidence** is important (why it matters)?	☐ No	☐ Almost	☐ Yes
6. Is the content of the argument **correct** based on the science concepts we talked about in class?	☐ No	☐ Almost	☐ Yes
Reviewers: If your group gave the author any "No" or "Almost" ratings, please give the author some advice about what to do to improve this part of his or her investigation report.			

Continued

Investigation Log

Section 3: Mechanics	Reviewer Rating		
1. **Grammar:** Are the sentences complete? Is there proper subject-verb agreement in each sentence? Are there no run-on sentences?	☐ No	☐ Almost	☐ Yes
2. **Conventions:** Did the author use proper spelling, punctuation, and capitalization?	☐ No	☐ Almost	☐ Yes
3. **Word Choice:** Did the author use the right words in each sentence (for example, *there* vs. *their, to* vs. *too, then* vs. *than*)?	☐ No	☐ Almost	☐ Yes

Reviewers: If your group gave the author any "No" or "Almost" ratings, please give the author some advice about what to do to improve the writing mechanics of his or her investigation report.

General Reviewer Comments

We liked …

We wonder …

Investigation 10. Patterns and the Transfer of Information: How Can We Use an Electric Circuit to Accurately Transfer Information About the Content of a Picture?

Write Your Final Report

Once you have received feedback from your friends about your draft report, create your final investigation report in the space that follows.

Introduction

_____
_____
_____
_____
_____
_____
_____
_____

Method

_____
_____
_____
_____
_____
_____
_____
_____
_____
_____
_____
_____
_____

Investigation Log

Argument

Investigation 10. Patterns and the Transfer of Information: How Can We Use an Electric Circuit to Accurately Transfer Information About the Content of a Picture?

Investigation Report Grading Rubric

Section 1: The Investigation	Score Missing	Somewhat	Yes
1. The author explained what the investigation was about.	0	1	2
2. The author made the **guiding question** clear.	0	1	2
3. The author **described** what he or she did to **collect data.**	0	1	2
4. The author described **how** he or she **analyzed** the data.	0	1	2
Section 2: The Argument	**Score Missing**	**Somewhat**	**Yes**
1. The claim includes a clear and detailed **answer** to the guiding question.	0	1	2
2. The author used **scientific evidence** to support the claim. Scientific evidence includes analyzed data and an explanation of the analysis.	0	1	2
3. The evidence **supports the claim.**	0	1	2
4. The author included enough **evidence** in his or her argument.	0	1	2
5. The author **explained why the evidence** is important.	0	1	2
6. The content of the argument is **correct.**	0	1	2
Section 3: Mechanics	**Score Missing**	**Somewhat**	**Yes**
1. *Grammar:* The sentences are complete. There is proper subject-verb agreement in each sentence. There are no run-on sentences.	0	1	2
2. *Conventions:* The author used proper spelling, punctuation, and capitalization.	0	1	2
3. *Word Choice:* The author used the right words in each sentence (e.g., *there* vs. *their, to* vs. *too, then* vs. *than*).	0	1	2

Teacher Comments

Here are some things I really liked about your report …	Here are some things I think you could do next time to make your report even better …

Total: _____ /26

Checkout Questions

Investigation 10. Patterns and the Transfer of Information

Draw a picture in the grid below by shading in some boxes and leaving some boxes blank.

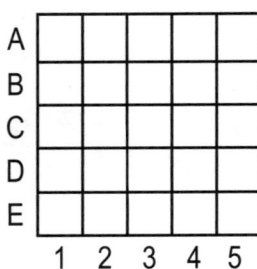

1. Imagine that a friend wants to see your picture but he or she lives far away. How would you code your picture so you could send it to your friend digitally?

   A1:   B1:   C1:   D1:   E1:

   A2:   B2:   C2:   D2:   E2:

   A3:   B3:   C3:   D3:   E3:

   A4:   B4:   C4:   D3:   E4:

   A5:   B5:   C5:   D4:   E5:

2. Explain your thinking. How can you use *patterns* to send people information?

   _____
   _____
   _____
   _____
   _____
   _____

Teacher Scoring Rubric for the Checkout Questions

Level	Description
3	The student can apply the core idea correctly in all cases and can fully explain the pattern.
2	The student can apply the core idea correctly in all cases but cannot fully explain the pattern.
1	The student cannot apply the core idea correctly in all cases but can fully explain the pattern.
0	The student cannot apply the core idea correctly in all cases and cannot explain the pattern.

National Science Teachers Association

Section 4
From Molecules to Organisms: Structures and Processes

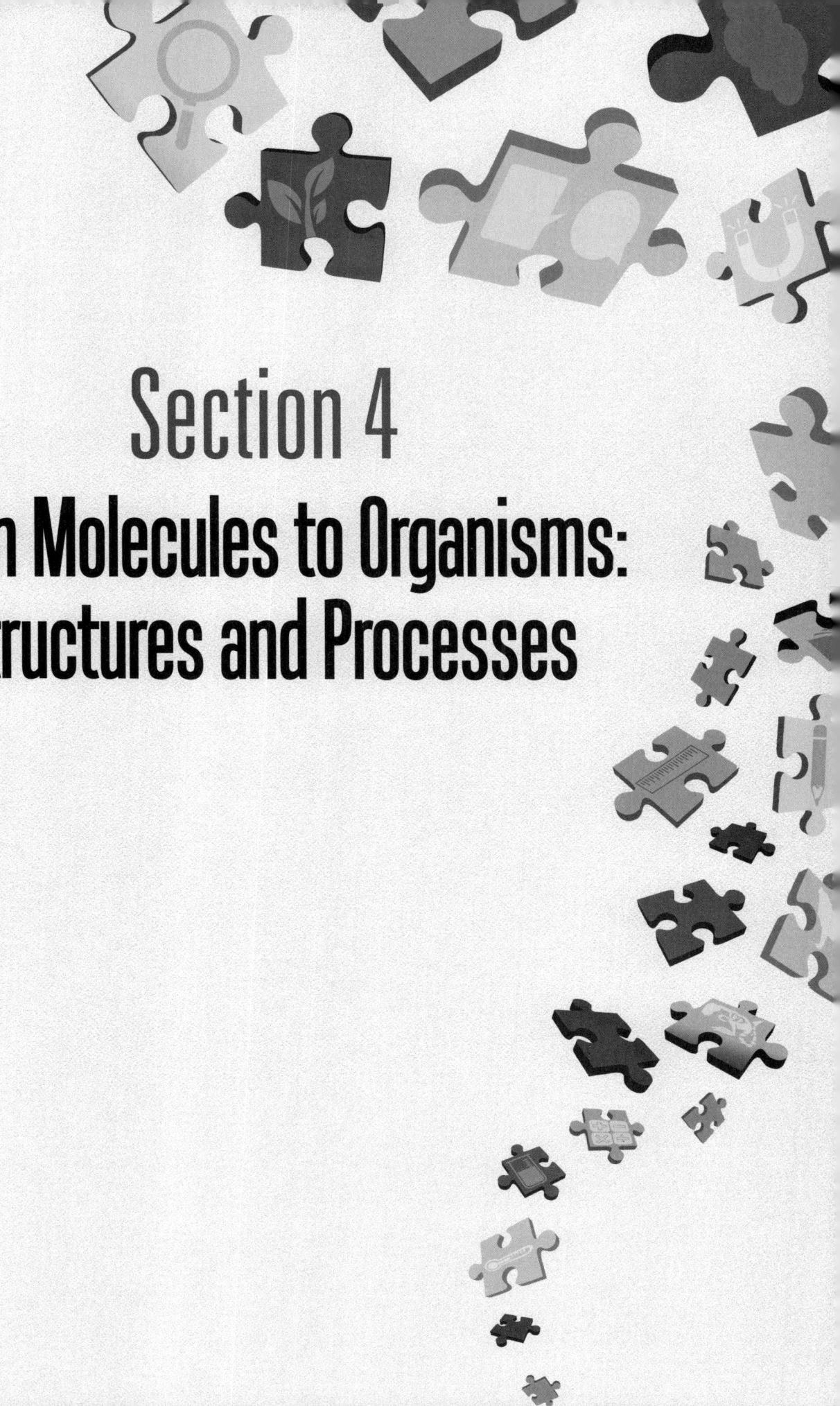

Investigation Log

Investigation 11

Structures of Plants: How Does Water Move From the Roots to the Leaves of a Plant?

Introduction

When you look at all the plants outside your home or school, you will probably notice a few things. First, there are many different kinds of plants. Second, many different types of plants share a lot of the same structures. Take a moment to examine a few different plants. Be sure to keep track of what you observe about the structures of these plants and what you are wondering about as you examine them in the boxes below.

Things I OBSERVED …	Things I WONDER about …

The external structures of a plant, which include such things as leaves, roots, stems, and flowers, have specific functions. Roots enable a plant to absorb minerals and water from the soil. They also keep the plant in one place so it does not blow away in the wind. The stem of a plant provides support for the leaves. Leaves absorb sunlight and carbon dioxide. Leaves are also where plants turn carbon dioxide

Investigation Log

and water into sugar. Plants use the sugar they make as a source of energy. The flowers of a plant are used for reproduction. All of these structures are important because they enable a plant to survive, grow, and reproduce.

There are also several structures found inside plants that are important. One such structure is used to transport water from the roots of the plant, through the stem, and up to the leaves. Without this internal structure, a plant would quickly die because it would not be able to make the sugar it needs.

Your goal in this investigation is to figure out how water travels from the roots of a plant to its leaves. To accomplish this goal, you will need to think of a plant as a system and then track the movement of water through that system over time. You will also need to be able to identify the internal structures that move water from the roots to the leaves. You can track the movement of water through a plant by placing the stem of a plant in water that is mixed with food coloring. The food coloring will move through the plant with the water and stain the parts of the plants that it touches (just like food coloring stains your skin when you touch it). You can also cut the stem of a plant lengthwise and place each half of the stem in a different cup of colored water and see what happens. You will also be able to cut open the plant after the food coloring has traveled through the plant so you can take a closer look at the inside of a stem. Once you learn more about the structures of the plant, and how these structures work together as a system, you can develop a model to explain how water travels from the roots, through the stem, and into the leaves of a plant.

Things we KNOW from what we read …	What we will NEED to figure out …

Investigation 11. Structures of Plants:
How Does Water Move From the Roots to the Leaves of a Plant?

Your Task

Use what you know about plants, structure and function, and the importance of tracking the movement of matter in a system to develop a model that explains how water moves through a plant.

The *guiding question* of this investigation is, **How does water move from the roots to the leaves of a plant?**

Materials

You may use any of the following materials during your investigation:

- Safety glasses or goggles (required)
- Apron (required)
- Celery
- Bok choy (Chinese cabbage)
- White carnations
- Food coloring (four different colors)
- Plastic cups
- Hand lens

Safety Rules

Follow all normal lab safety rules. In addition, be sure to follow these rules:

- Wear sanitized safety glasses or goggles and a nonlatex apron during setup, investigation activity, and cleanup.
- Immediately clean up any spills to avoid a slip or fall hazard.
- Do not eat or drink any food items used in an investigation activity.
- Ask your teacher to cut the plants when you need them cut.
- Wash your hands with soap and water when you are done collecting the data.

Plan Your Investigation

Prepare a plan for your investigation by filling out the chart that follows; this plan is called an *investigation proposal*. Before you start developing your plan, be sure to discuss the following questions with the other members of your group:

- What parts of the **system** are you studying?
- How can we **track the movement of matter within this system**?

Investigation Log

Our guiding question:

This is a picture of how we will set up the equipment:

We will collect the following data:

These are the steps we will follow to collect data:

I approve of this investigation proposal.

_____   _____
Teacher's signature   Date

Investigation 11. Structures of Plants:
How Does Water Move From the Roots to the Leaves of a Plant?

Collect Your Data

Keep a record of what you measure or observe during your investigation in the space below.

Analyze Your Data

You will need to analyze the data you collected before you can develop an answer to the guiding question. In the space below, create a table, graph, or one or more pictures.

Investigation Log

Draft Argument

Develop an argument on a whiteboard. It should include the following:

- A *claim*: Your answer to the guiding question.
- *Evidence*: An analysis of the data and an explanation of what the analysis means.
- A *justification of the evidence*: Why your group thinks the evidence is important.

The Guiding Question:	
Our Claim:	
Our Evidence:	Our Justification of the Evidence:

Argumentation Session

Share your argument with your classmates. Be sure to ask them how to make your draft argument better. Keep track of their suggestions in the space below.

Ways to IMPROVE our argument …

Investigation 11. Structures of Plants:
How Does Water Move From the Roots to the Leaves of a Plant?

Draft Report

Prepare an *investigation report* to share what you have learned. Use the information in this handout and your group's final argument to write a *draft* of your investigation report.

Introduction

We have been studying _____ in class. Before we

started this investigation, we explored _____

_____

_____

We noticed _____

_____

_____

My goal for this investigation was to figure out _____

_____

_____

The guiding question was _____

_____

_____

Method

To gather the data I needed to answer this question, I _____

_____

_____

_____

_____

Student Workbook for Argument-Driven Inquiry in **Fourth-Grade Science:** Three-Dimensional Investigations

Investigation Log

I then analyzed the data I collected by _____

_____

_____

_____

_____

_____

Argument

My claim is _____

_____

_____

_____

_____

The information below shows _____

_____

_____

_____

National Science Teachers Association

Investigation 11. Structures of Plants:
How Does Water Move From the Roots to the Leaves of a Plant?

This analysis of the data I collected suggests _____

_____

_____

_____

_____

_____

This evidence is based on several important scientific concepts. The first one is _____

_____

_____

_____

_____

_____

 **Review**

Your friends need your help! Review the draft of their investigation reports and give them ideas about how to improve. Use the *peer-review guide* that begins on the next page to guide your review.

Investigation Log

Peer-Review Guide

Section 1: The Investigation		Reviewer Rating	
1. Did the author do a good job of explaining what the investigation was about?	☐ No	☐ Almost	☐ Yes
2. Did the author do a good job of making the **guiding question** clear?	☐ No	☐ Almost	☐ Yes
3. Did the author do a good job of describing what he or she did to **collect data?**	☐ No	☐ Almost	☐ Yes
4. Did the author do a good job describing **how** he or she **analyzed** the data?	☐ No	☐ Almost	☐ Yes

Reviewers: If your group gave the author any "No" or "Almost" ratings, please give the author some advice about what to do to improve this part of his or her investigation report.

Section 2: The Argument		Reviewer Rating	
1. Does the author's claim provide a clear and detailed **answer** to the guiding question?	☐ No	☐ Almost	☐ Yes
2. Did the author support his or her claim with **scientific evidence?** Scientific evidence includes analyzed data and an explanation of the analysis.	☐ No	☐ Almost	☐ Yes
3. Does the **evidence** that the author uses in his or her argument **support the claim?**	☐ No	☐ Almost	☐ Yes
4. Did the author include enough **evidence** in his or her argument?	☐ No	☐ Almost	☐ Yes
5. Did the author do a good job of **explaining why the evidence** is important (why it matters)?	☐ No	☐ Almost	☐ Yes
6. Is the content of the argument **correct** based on the science concepts we talked about in class?	☐ No	☐ Almost	☐ Yes

Reviewers: If your group gave the author any "No" or "Almost" ratings, please give the author some advice about what to do to improve this part of his or her investigation report.

Continued

Investigation 11. Structures of Plants:
How Does Water Move From the Roots to the Leaves of a Plant?

Section 3: Mechanics	Reviewer Rating		
1. ***Grammar:*** Are the sentences complete? Is there proper subject-verb agreement in each sentence? Are there no run-on sentences?	☐ No	☐ Almost	☐ Yes
2. ***Conventions:*** Did the author use proper spelling, punctuation, and capitalization?	☐ No	☐ Almost	☐ Yes
3. ***Word Choice:*** Did the author use the right words in each sentence (for example, *there* vs. *their, to* vs. *too, then* vs. *than*)?	☐ No	☐ Almost	☐ Yes

Reviewers: If your group gave the author any "No" or "Almost" ratings, please give the author some advice about what to do to improve the writing mechanics of his or her investigation report.

General Reviewer Comments

We liked …

We wonder …

Investigation Log

Write Your Final Report

Once you have received feedback from your friends about your draft report, create your final investigation report in the space that follows.

Introduction

_____
_____
_____
_____
_____
_____
_____

Method

_____
_____
_____
_____
_____
_____
_____
_____
_____
_____
_____

Investigation 11. Structures of Plants:
How Does Water Move From the Roots to the Leaves of a Plant?

Argument

Investigation Log

Investigation Report Grading Rubric

Section 1: The Investigation	Missing	Somewhat	Yes
1. The author explained what the investigation was about.	0	1	2
2. The author made the **guiding question** clear.	0	1	2
3. The author **described** what he or she did to **collect data**.	0	1	2
4. The author described **how** he or she **analyzed** the data.	0	1	2

Section 2: The Argument	Missing	Somewhat	Yes
1. The claim includes a clear and detailed **answer** to the guiding question.	0	1	2
2. The author used **scientific evidence** to support the claim. Scientific evidence includes analyzed data and an explanation of the analysis.	0	1	2
3. The evidence **supports the claim.**	0	1	2
4. The author included enough **evidence** in his or her argument.	0	1	2
5. The author **explained why the evidence** is important.	0	1	2
6. The content of the argument is **correct**.	0	1	2

Section 3: Mechanics	Missing	Somewhat	Yes
1. *Grammar:* The sentences are complete. There is proper subject-verb agreement in each sentence. There are no run-on sentences.	0	1	2
2. *Conventions:* The author used proper spelling, punctuation, and capitalization.	0	1	2
3. *Word Choice:* The author used the right words in each sentence (e.g., *there* vs. *their*, *to* vs. *too*, *then* vs. *than*).	0	1	2

Teacher Comments

Here are some things I really liked about your report …	Here are some things I think you could do next time to make your report even better …

Total: _____ /26

Checkout Questions

Investigation 11. Structures of Plants

1. We can think of a plant as a system that includes several different structures. Create a model that shows how water moves into, out of, and within this system. Your model should include (1) the internal structures of the plant and how they function, (2) where the water comes from, (3) where the water goes, and (4) labels that describe what is happening in the model.

2. Explain your thinking. How the do the internal *structures* of a plant enable it to *function*?

___

Teacher Scoring Rubric for the Checkout Questions

Level	Description
3	The student can apply the core idea correctly in all cases and can fully explain how structure and function are related.
2	The student can apply the core idea correctly in all cases but cannot fully explain how structure and function are related.
1	The student cannot apply the core idea correctly in all cases but can fully explain how structure and function are related.
0	The student cannot apply the core idea correctly in all cases and cannot explain how structure and function are related.

Investigation Log

Investigation 12

Structures of Animals: How Should We Classify the Unknown Organisms?

Introduction

When you look at the animals living outside your home or school, you will probably notice a few things. First, there are many different kinds of animals. Second, many different types of animals share a lot of the same traits. For example, all cows and wolves have one tail, two eyes, and four legs. These animals are also covered in hair and have a mouth full of teeth. Although different types of animals have a lot in common, they also have specific structures that make them unique. Some of the structures of animals that make them unique are the location of their eyes, the size of their jaws, and the shape of their teeth. Take a moment to examine the eyes, jaw, and teeth of an animal. Be sure to keep track of what you observe and what you are wondering about in the boxes below as you examine the animal's skull.

Things I OBSERVED …	Things I WONDER about …

The structures of an animal, such as the location of its eyes, the size of its jaw, and the shape of its teeth, serve specific functions that enable the animal to survive and reproduce. For example, all animals

Investigation Log

must find and eat food to survive. Different types of animals need to find and eat different kinds of food. Some animals eat only plants. These animals are called *herbivores*. Some animals eat only meat. These animals are called *carnivores*. Some animals eat both plants and animals. These animals are called *omnivores*. Carnivores hunt and eat herbivores. Carnivores therefore need to be able to find and catch herbivores to survive. Herbivores, in contrast, need to be able to see and avoid carnivores. Carnivores have eyes located to make it easier for them to find and catch herbivores, and herbivores have eyes located to make it easier for them to see and avoid carnivores. Carnivores and herbivores also have teeth and a jaw shaped in a way that makes it easier for them to eat a specific kind of food.

Your goal in this investigation is to figure out how to classify an unknown animal based on the type of food it eats. The unknown organism can be a carnivore, an herbivore, or an omnivore. To accomplish this task, you will need to make observations and take measurements about the location of the eyes, the size of the jaw, and the shape of the teeth for six different animals: two carnivores, two herbivores, and two omnivores. You will need to compare and contrast these traits to figure out what these different types of animals have in common and what is different about them. Scientists often look for patterns in nature like this and then use these patterns to classify animals into groups. You can therefore use patterns to help classify the unknown organisms as carnivore, herbivore, or omnivore.

Things we KNOW from what we read …	What we will NEED to figure out …

Investigation 12. Structures of Animals:
How Should We Classify the Unknown Organisms?

Your Task

Use what you know about traits, the relationship between structure and function, and patterns to design and carry out an investigation to classify two unknown animals based on what they eat.

The *guiding question* of this investigation is, ***How should we classify the unknown organisms?***

Materials

You may use any of the following materials during your investigation:

- Skull A (herbivore)
- Skull B (herbivore)
- Skull C (carnivore)
- Skull D (carnivore)
- Skull E (omnivore)
- Skull F (omnivore)
- Unknown skull G
- Unknown skull H
- Ruler
- Protractor
- String or tape measure

Safety Rules

Follow all normal safety rules. In addition, be sure to follow these rules:

- Be careful when handling the skulls—they are fragile and can break, and the teeth can be sharp.
- Wash your hands with soap and water when you are done collecting the data.

Plan Your Investigation

Prepare a plan for your investigation by filling out the chart that follows; this plan is called an *investigation proposal*. Before you start developing your plan, be sure to discuss the following questions with the other members of your group:

- What types of **patterns** might we look for to help answer the guiding question?
- How might the **structure** of a skull be related to what an animal eats?

Student Workbook for Argument-Driven Inquiry in **Fourth-Grade Science**: Three-Dimensional Investigations

Investigation Log

Our guiding question:

We will collect the following data:

These are the steps we will follow to collect data:

I approve of this investigation proposal.

_____   _____
Teacher's signature   Date

National Science Teachers Association

Investigation 12. Structures of Animals:
How Should We Classify the Unknown Organisms?

Collect Your Data

Keep a record of what you measure or observe during your investigation in the space below.

Analyze Your Data

You will need to analyze the data you collected before you can develop an answer to the guiding question. In the space below, create a table, a graph, or pictures.

Investigation Log

Draft Argument

Develop an argument on a whiteboard. It should include the following:

- A *claim*: Your answer to the guiding question.
- *Evidence*: An analysis of the data and an explanation of what the analysis means.
- A *justification of the evidence*: Why your group thinks the evidence is important.

The Guiding Question:	
Our Claim:	
Our Evidence:	Our Justification of the Evidence:

Argumentation Session

Share your argument with your classmates. Be sure to ask them how to make your draft argument better. Keep track of their suggestions in the space below.

Ways to IMPROVE our argument …

Investigation 12. Structures of Animals:
How Should We Classify the Unknown Organisms?

 **Draft Report**

Prepare an *investigation report* to share what you have learned. Use the information in this handout and your group's final argument to write a *draft* of your investigation report.

Introduction

We have been studying _____ in class. Before we started

this investigation, we explored _____

_____

_____

We noticed _____

_____

_____

My goal for this investigation was to figure out _____

_____

_____

The guiding question was _____

_____

_____

Method

To gather the data I needed to answer this question, I _____

_____

_____

_____

I then analyzed the data I collected by _____

_____

_____

_____

Investigation Log

Argument

My claim is _____

_____

_____

_____

_____

The information below shows _____

_____

_____

This analysis of the data I collected suggests _____

_____

_____

_____

_____

Investigation 12. Structures of Animals:
How Should We Classify the Unknown Organisms?

This evidence is important because of several scientific concepts. The first one is _____

_____

_____

_____

_____

_____

_____

 **Review**

Your friends need your help! Review the draft of their investigation reports and give them ideas about how to improve. Use the *peer-review guide* that begins on the next page to guide your review.

Peer-Review Guide

Section 1: The Investigation		Reviewer Rating	
1. Did the author do a good job of explaining what the investigation was about?	☐ No	☐ Almost	☐ Yes
2. Did the author do a good job of making the **guiding question** clear?	☐ No	☐ Almost	☐ Yes
3. Did the author do a good job of describing what he or she did to **collect data**?	☐ No	☐ Almost	☐ Yes
4. Did the author do a good job describing **how** he or she **analyzed** the data?	☐ No	☐ Almost	☐ Yes

Reviewers: If your group gave the author any "No" or "Almost" ratings, please give the author some advice about what to do to improve this part of his or her investigation report.

Section 2: The Argument		Reviewer Rating	
1. Does the author's claim provide a clear and detailed **answer** to the guiding question?	☐ No	☐ Almost	☐ Yes
2. Did the author support his or her claim with **scientific evidence**? Scientific evidence includes analyzed data and an explanation of the analysis.	☐ No	☐ Almost	☐ Yes
3. Does the **evidence** that the author uses in his or her argument **support the claim**?	☐ No	☐ Almost	☐ Yes
4. Did the author include enough **evidence** in his or her argument?	☐ No	☐ Almost	☐ Yes
5. Did the author do a good job of **explaining why the evidence** is important (why it matters)?	☐ No	☐ Almost	☐ Yes
6. Is the content of the argument **correct** based on the science concepts we talked about in class?	☐ No	☐ Almost	☐ Yes

Reviewers: If your group gave the author any "No" or "Almost" ratings, please give the author some advice about what to do to improve this part of his or her investigation report.

Continued

Investigation 12. Structures of Animals:
How Should We Classify the Unknown Organisms?

Section 3: Mechanics	Reviewer Rating		
1. *Grammar:* Are the sentences complete? Is there proper subject-verb agreement in each sentence? Are there no run-on sentences?	☐ No	☐ Almost	☐ Yes
2. *Conventions:* Did the author use proper spelling, punctuation, and capitalization?	☐ No	☐ Almost	☐ Yes
3. *Word Choice:* Did the author use the right words in each sentence (for example, *there* vs. *their, to* vs. *too, then* vs. *than*)?	☐ No	☐ Almost	☐ Yes

Reviewers: If your group gave the author any "No" or "Almost" ratings, please give the author some advice about what to do to improve the writing mechanics of his or her investigation report.

General Reviewer Comments

We liked …

We wonder …

Investigation Log

Write Your Final Report

Once you have received feedback from your friends about your draft report, create your final investigation report in the space that follows.

Introduction

_____
_____
_____
_____
_____
_____
_____

Method

_____
_____
_____
_____
_____
_____
_____
_____
_____
_____
_____
_____

Investigation 12. Structures of Animals:
How Should We Classify the Unknown Organisms?

Argument

Investigation Log

Investigation Report Grading Rubric

	Score Missing	Somewhat	Yes
Section 1: The Investigation			
1. The author explained what the investigation was about.	0	1	2
2. The author made the **guiding question** clear.	0	1	2
3. The author **described** what he or she did to **collect data.**	0	1	2
4. The author described **how** he or she **analyzed** the data.	0	1	2
Section 2: The Argument	**Score** Missing	Somewhat	Yes
1. The claim includes a clear and detailed **answer** to the guiding question.	0	1	2
2. The author used **scientific evidence** to support the claim. Scientific evidence includes analyzed data and an explanation of the analysis.	0	1	2
3. The evidence **supports the claim.**	0	1	2
4. The author included enough **evidence** in his or her argument.	0	1	2
5. The author **explained why the evidence** is important.	0	1	2
6. The content of the argument is **correct.**	0	1	2
Section 3: Mechanics	**Score** Missing	Somewhat	Yes
1. *Grammar:* The sentences are complete. There is proper subject-verb agreement in each sentence. There are no run-on sentences.	0	1	2
2. *Conventions:* The author used proper spelling, punctuation, and capitalization.	0	1	2
3. *Word Choice:* The author used the right words in each sentence (e.g., *there* vs. *their*, *to* vs. *too*, *then* vs. *than*).	0	1	2

Teacher Comments

Here are some things I really liked about your report …	Here are some things I think you could do next time to make your report even better …

Total: _____ /26

Checkout Questions

Investigation 12. Structures of Animals

1. Pictured below is the skull of an animal. What type of food do you think this animal would have eaten when it was alive?

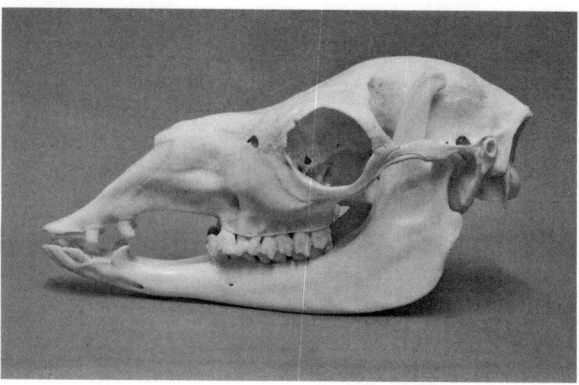

☐ Animals only

☐ Plants only

☐ Both plants and animals

2. Pictured below is the skull of an animal. What type of food do you think this animal would have eaten when it was alive?

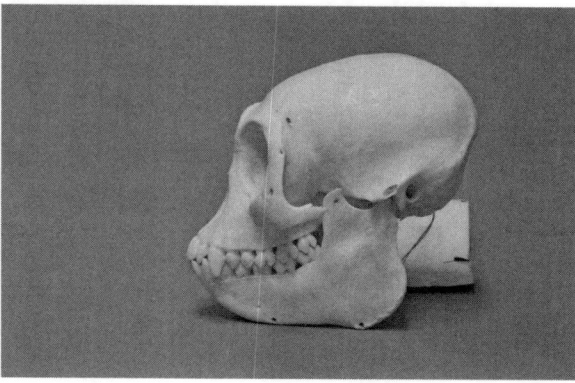

☐ Animals only

☐ Plants only

☐ Both plants and animals

Checkout Questions

3. Pictured below is the skull of an animal. What type of food do you think this animal would have eaten when it was alive?

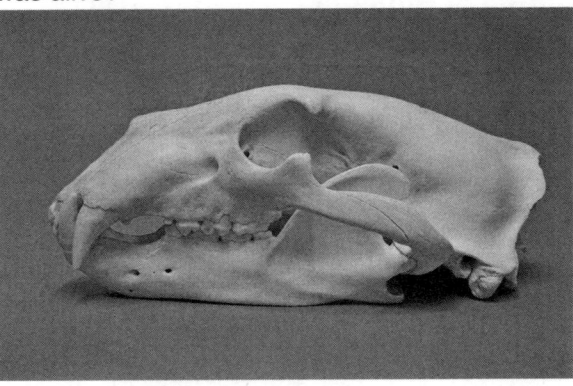

☐ Animals only

☐ Plants only

☐ Both plants and animals

4. Explain your thinking. How did the *structure* of the skulls of these animals allow you to determine what they would have eaten when they were alive?

_____
_____
_____
_____
_____
_____
_____

Teacher Scoring Rubric for the Checkout Questions

Level	Description
3	The student can apply the core idea correctly in all cases and can fully explain how structure and function are related.
2	The student can apply the core idea correctly in all cases but cannot fully explain how structure and function are related.
1	The student cannot apply the core idea correctly in all cases but can fully explain how structure and function are related.
0	The student cannot apply the core idea correctly in all cases and cannot explain how structure and function are related.

Investigation Log

Investigation 13

Information From Senses: What Type of Environmental Conditions Do Mealworms Seek Out?

Introduction

The mealworm is not a worm. It is a darkling beetle in the larva stage of its life cycle. Take a few minutes to examine a mealworm up close. Be sure to keep track of what you observe and what you are wondering about in the boxes below.

Things I OBSERVED …	Things I WONDER about …

People will often find a few mealworms in a box of cereal, flour, or oatmeal that has been left open in the pantry or cupboard, because mealworms eat grains. People will sometimes find mealworms in other locations as well. For example, a property owner posted the following comments to an online discussion board after he found mealworms at one of his rental properties:

> *I found hundreds of mealworms in the refrigerator of a cabin that I rent out to people. The refrigerator wasn't cleaned out before the power was turned off. The door of the refrigerator was left open as well. I also found what I would "guess" is excrement in large quantities near an open bag of oatmeal that was left on the floor. There are trails between the bag of oatmeal and the fridge. I need to get this cabin ready to rent again so I have to make sure that I get rid of all of the mealworms. I just don't know where else I should look for them. I would appreciate some advice about what to do next.*

Investigation Log

A mealworm, like all animals, seeks out environmental conditions that will enable it to find what it needs to survive, grow, or reproduce. To find a suitable environment, an animal must be able to gather information about its surroundings. Animals rely on their senses to collect this information. For example, mealworms can smell different odors, feel different textures, hear different sound waves, and taste different flavors. Animals process this information based on their current needs and, in some cases, their memories of past experiences. Once the information is processed, the animal will then respond in a way that will help it survive, grow, or reproduce. A mealworm, for example, might move toward an object because the odor of the object is associated with a type of food it eats, or it might stop eating something that has a specific taste because that taste is associated with something that is deadly. The sense receptors, brain, and body of an animal therefore function together as a system. This system allows an animal to receive, process, and act on information in a way that enables it to find what it needs and avoid things that are harmful.

Your goal in this investigation is to figure out the types of environmental conditions that mealworms seek out. To accomplish this goal, you will need to look for patterns in the ways that mealworms react to different environmental conditions. Your teacher will show you how to set up and use a choice chamber to expose mealworms to two different environmental conditions at the same time. As you design and carry out your investigation, be sure to keep in mind that mealworms collect information about the environment using their senses, process this information with their brain, and then react in a way that will help them survive, grow, or reproduce.

Things we KNOW from what we read …	What we will NEED to figure out …

Investigation 13. Information From Senses:
What Type of Environmental Conditions Do Mealworms Seek Out?

Your Task

Use what you know about how animals process information, systems, and patterns to design and carry out an investigation to figure out how mealworms respond to different types of environmental conditions.

The *guiding question* of this investigation is, **What type of environmental conditions do mealworms seek out?**

Materials

You may use any of the following materials during your investigation:

- Safety glasses or goggles (required)
- Nonlatex gloves (required)
- 18 mealworms
- Oatmeal
- 3 two-compartment petri dishes
- Water
- Pipette
- Filter paper
- Black construction paper
- Tape
- Stopwatch
- Hand lens
- Stirrer

Safety Rules

Follow all normal safety rules. In addition, be sure to follow these rules:

- Wear sanitized safety glasses or goggles and nonlatex gloves during setup, investigation activity, and cleanup.
- Be careful with the mealworms. Keep them healthy by not doing anything to hurt them such as pulling on them, poking them, cutting them, or dropping them.
- Do not eat the oatmeal.
- Wash your hands with soap and water when you are done collecting the data.

Plan Your Investigation

Prepare a plan for your investigation by filling out the chart that follows; this plan is called an *investigation proposal*. Before you start developing your plan, be sure to discuss the following questions with the other members of your group:

- What part or component of the mealworm body **system** are you studying?
- What types of **patterns** might we look for to help answer the guiding question?

Investigation Log

Our guiding question:

This is a picture of how we will set up the equipment:

We will collect the following data:

These are the steps we will follow to collect data:

I approve of this investigation proposal.

_____   _____
Teacher's signature   Date

Investigation 13. Information From Senses:
What Type of Environmental Conditions Do Mealworms Seek Out?

Collect Your Data

Keep a record of what you measure or observe during your investigation in the space below.

Analyze Your Data

You will need to analyze the data you collected before you can develop an answer to the guiding question. To do this, create a graph that shows a pattern or a relationship.

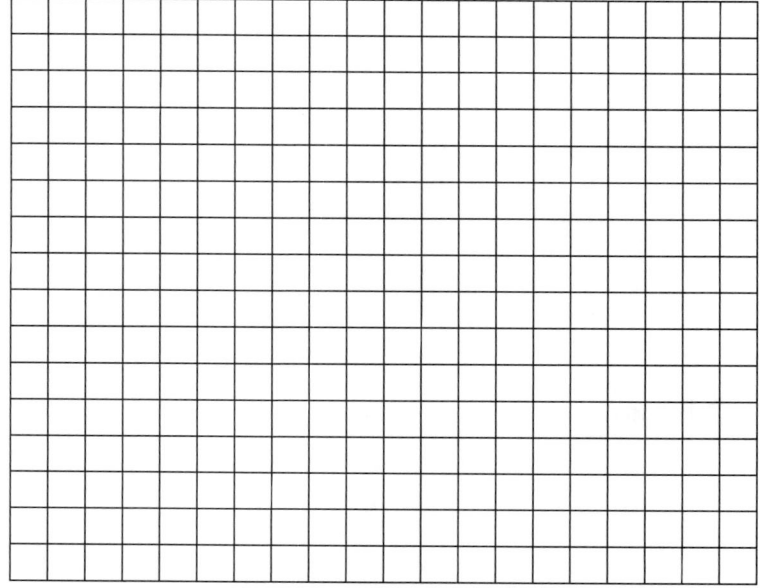

Investigation Log

Draft Argument

Develop an argument on a whiteboard. It should include the following:

- A *claim*: Your answer to the guiding question.
- *Evidence*: An analysis of the data and an explanation of what the analysis means.
- A *justification of the evidence*: Why your group thinks the evidence is important.

The Guiding Question:	
Our Claim:	
Our Evidence:	Our Justification of the Evidence:

Argumentation Session

Share your argument with your classmates. Be sure to ask them how to make your draft argument better. Keep track of their suggestions in the space below.

Ways to IMPROVE our argument …

Draft Report

Prepare an *investigation report* to share what you have learned. Use the information in this handout and your group's final argument to write a draft of your investigation report.

Investigation 13. Information From Senses:
What Type of Environmental Conditions Do Mealworms Seek Out?

Introduction

We have been studying _____ in class. Before we

started this investigation, we explored _____

_____

_____

We noticed _____

_____

_____

My goal for this investigation was to figure out _____

_____

_____

The guiding question was _____

_____

_____

Method

To gather the data I needed to answer this question, I _____

_____

_____

_____

_____

_____

_____

_____

Investigation Log

I then analyzed the data I collected by _____
_____
_____
_____
_____

Argument

My claim is _____
_____
_____
_____

The graph below shows _____
_____
_____
_____
_____

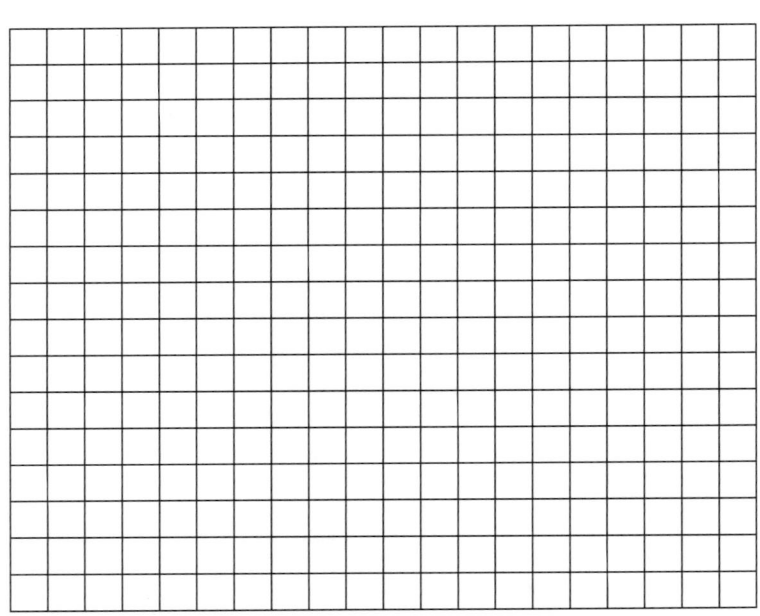

Investigation 13. Information From Senses:
What Type of Environmental Conditions Do Mealworms Seek Out?

This analysis of the data I collected suggests _____

_____

_____

_____

_____

_____

_____

This evidence is important because of several scientific concepts. The first one is _____

_____

_____

_____

_____

_____

_____

_____

_____

_____

 **Review**

Your friends need your help! Review the draft of their investigation reports and give them ideas about how to improve. Use the *peer-review guide* that begins on the next page to guide your review.

Investigation Log

Peer-Review Guide

Section 1: The Investigation		Reviewer Rating	
1. Did the author do a good job of explaining what the investigation was about?	☐ No	☐ Almost	☐ Yes
2. Did the author do a good job of making the **guiding question** clear?	☐ No	☐ Almost	☐ Yes
3. Did the author do a good job of describing what he or she did to **collect data**?	☐ No	☐ Almost	☐ Yes
4. Did the author do a good job describing **how** he or she **analyzed** the data?	☐ No	☐ Almost	☐ Yes
Reviewers: If your group gave the author any "No" or "Almost" ratings, please give the author some advice about what to do to improve this part of his or her investigation report.			

Section 2: The Argument		Reviewer Rating	
1. Does the author's claim provide a clear and detailed **answer** to the guiding question?	☐ No	☐ Almost	☐ Yes
2. Did the author support his or her claim with **scientific evidence?** Scientific evidence includes analyzed data and an explanation of the analysis.	☐ No	☐ Almost	☐ Yes
3. Does the **evidence** that the author uses in his or her argument **support the claim?**	☐ No	☐ Almost	☐ Yes
4. Did the author include enough **evidence** in his or her argument?	☐ No	☐ Almost	☐ Yes
5. Did the author do a good job of **explaining why the evidence** is important (why it matters)?	☐ No	☐ Almost	☐ Yes
6. Is the content of the argument **correct** based on the science concepts we talked about in class?	☐ No	☐ Almost	☐ Yes
Reviewers: If your group gave the author any "No" or "Almost" ratings, please give the author some advice about what to do to improve this part of his or her investigation report.			

Continued

Investigation 13. Information From Senses:
What Type of Environmental Conditions Do Mealworms Seek Out?

Section 3: Mechanics	Reviewer Rating		
1. ***Grammar:*** Are the sentences complete? Is there proper subject-verb agreement in each sentence? Are there no run-on sentences?	☐ No	☐ Almost	☐ Yes
2. ***Conventions:*** Did the author use proper spelling, punctuation, and capitalization?	☐ No	☐ Almost	☐ Yes
3. ***Word Choice:*** Did the author use the right words in each sentence (for example, *there* vs. *their, to* vs. *too, then* vs. *than*)?	☐ No	☐ Almost	☐ Yes

Reviewers: If your group gave the author any "No" or "Almost" ratings, please give the author some advice about what to do to improve the writing mechanics of his or her investigation report.

General Reviewer Comments

We liked …

We wonder …

Investigation Log

Write Your Final Report

Once you have received feedback from your friends about your draft report, create your final investigation report in the space that follows.

Introduction

Method

Investigation 13. Information From Senses:
What Type of Environmental Conditions Do Mealworms Seek Out?

Argument

Investigation Log

Investigation Report Grading Rubric

Section 1: The Investigation	Score Missing	Somewhat	Yes
1. The author explained what the investigation was about.	0	1	2
2. The author made the **guiding question** clear.	0	1	2
3. The author **described** what he or she did to **collect data.**	0	1	2
4. The author described **how** he or she **analyzed** the data.	0	1	2

Section 2: The Argument	Score Missing	Somewhat	Yes
1. The claim includes a clear and detailed **answer** to the guiding question.	0	1	2
2. The author used **scientific evidence** to support the claim. Scientific evidence includes analyzed data and an explanation of the analysis.	0	1	2
3. The evidence **supports the claim.**	0	1	2
4. The author included enough **evidence** in his or her argument.	0	1	2
5. The author **explained why the evidence** is important.	0	1	2
6. The content of the argument is **correct.**	0	1	2

Section 3: Mechanics	Score Missing	Somewhat	Yes
1. **Grammar:** The sentences are complete. There is proper subject-verb agreement in each sentence. There are no run-on sentences.	0	1	2
2. **Conventions:** The author used proper spelling, punctuation, and capitalization.	0	1	2
3. **Word Choice:** The author used the right words in each sentence (e.g., *there* vs. *their, to* vs. *too, then* vs. *than*).	0	1	2

Teacher Comments

Here are some things I really liked about your report …	Here are some things I think you could do next time to make your report even better …

Total: _____ /26

Checkout Questions

Investigation 13. Information From Senses

Use the following information to answer questions 1–3. Pictured below are the results of an experiment that was carried out to determine the needs of mealworms. Scientists put 12 mealworms in a container that was divided into four different sections. At the beginning of the experiment, there were 3 mealworms in each section. After 60 minutes, the number of mealworms in each section was different.

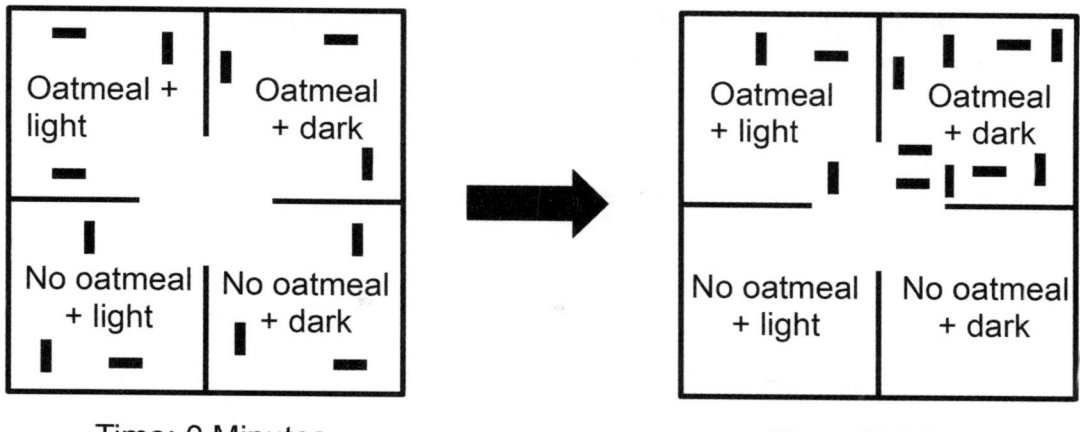

Time: 0 Minutes　　　　　　　　　　　Time: 60 Minutes

The mealworms moved around inside the container during the experiment because the environment in the four sections of the container was not the same.

1. How are mealworms able to detect, process, and use information about the environment around them?

_____
_____
_____
_____
_____
_____
_____
_____
_____

Student Workbook for Argument-Driven Inquiry in **Fourth-Grade Science**: Three-Dimensional Investigations

Checkout Questions

2. What conclusion can you draw from these results?

    a. Mealworms seek out food and light.

    b. Mealworms seek out food and darkness.

    c. Mealworms seek out light but not food.

    d. Mealworms seek out darkness but not food.

3. Explain your thinking. How did you use what you know about *patterns* to answer these questions?

    _____
    _____
    _____
    _____
    _____
    _____
    _____
    _____
    _____
    _____
    _____

Teacher Scoring Rubric for the Checkout Questions

Level	Description
3	The student can apply the core idea correctly in all cases and can fully explain the pattern.
2	The student can apply the core idea correctly in all cases but cannot fully explain the pattern.
1	The student cannot apply the core idea correctly in all cases but can fully explain the pattern.
0	The student cannot apply the core idea correctly in all cases and cannot explain the pattern.

Section 5
Earth's Place in the Universe and Systems

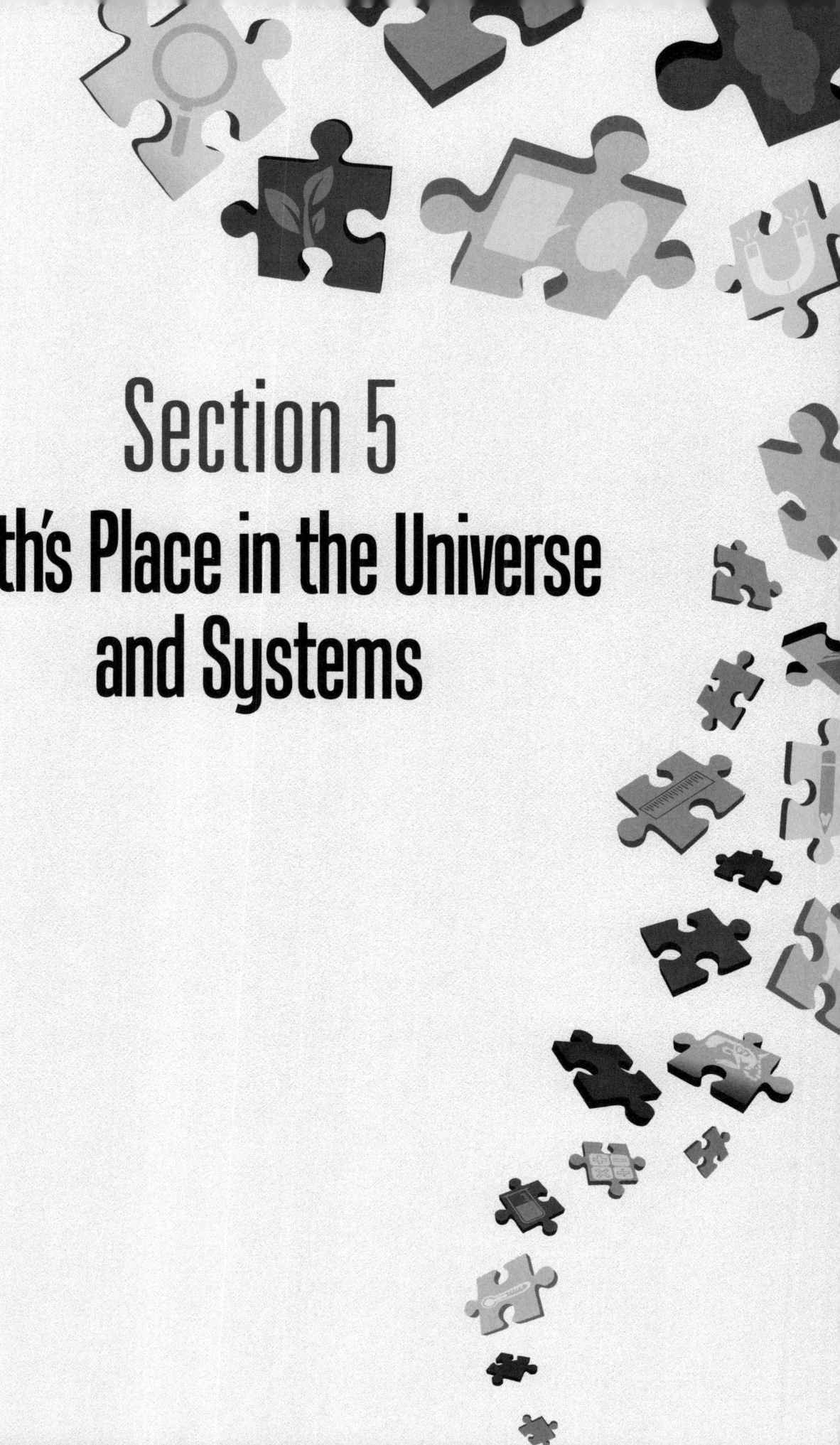

Investigation Log

Investigation 14

Movement of Water: Why Can We See the Roots of Trees That Grow Near Rivers or Streams?

Introduction

Trees can be found all over the world. A tree is a plant with a trunk made of wood. The trunk of a tree supports branches that hold leaves above the ground. A tree also has roots. The roots of a tree anchor it to the ground. The roots also make it possible for the tree to collect water and nutrients from the soil around it. Your teacher will have you look at some pictures of trees. Keep track of what you observe when you look at these trees and what you are wondering about as you look at them in the boxes below.

Things I OBSERVED ...	Things I WONDER about ...

Student Workbook for Argument-Driven Inquiry in **Fourth-Grade Science**: Three-Dimensional Investigations

Investigation Log

We usually cannot see the roots of trees because they are underground. However, some of the trees that you looked at have roots that you can see. These trees grow near a body of water such as a lake, river, or stream. Trees need water to survive, and the amount of water available at a particular location determines how many and what types of trees are found at that location.

The water found in lakes, rivers, and streams comes from rain and snow. The amount of rain or snow that falls in an area determines the size of the lakes, rivers, and streams in the area. When there is a lot of rain or snow, lakes, rivers, and streams will fill with water and grow bigger and deeper. When there is very little rain or snow, the lakes, rivers, and streams in the area will shrink in size and may even disappear. Water moving in a river or stream can also break rocks and soils into smaller pieces and move the smaller pieces from one place to another. Water can therefore change the appearance of the land in a region.

In this investigation, your goal is to figure out why we often see the roots of trees that are found growing near a river or stream. You will need to create a physical model of a river with trees along it for this investigation. Your teacher will show you how to create your model using a stream table, sand, some pipe cleaners, and water. You can then use your physical model to test out your different ideas about what you think happens to the trees that grow near rivers and streams.

Things we KNOW from what we read …	What we will NEED to figure out …

Investigation 14. Movement of Water:
Why Can We See the Roots of Trees That Grow Near Rivers or Streams?

Your Task

Use what you know about the movement of water over land, how water can move things around, and cause-and-effect relationships to design and carry out an investigation to determine how the movement of water in a river or stream (a *cause*) affects the soil around a tree (an *effect*).

The *guiding question* of this investigation is, **Why can we see the roots of trees that grow near rivers or streams?**

Materials

You may use any of the following materials during your investigation:

- Safety glasses or goggles (required)
- Plant tray
- Small block of wood
- 10 pipe cleaners
- Funnel pitcher
- Rulers
- Stopwatch
- 1 cm grid transparency
- Wet-erase markers
- Water
- Sand

Safety Rules

Follow all normal safety rules. In addition, be sure to follow these rules:

- Wear sanitized safety glasses or goggles during setup, investigation activity, and cleanup.
- Do not throw pipe cleaners or sand, and do not put these materials in your mouth.
- Immediately clean up any spills to avoid a slip or fall hazard.
- Wash your hands with soap and water when you are done collecting the data.

Plan Your Investigation

Prepare a plan for your investigation by filling out the chart that follows; this plan is called an *investigation proposal*. Before you start developing your plan, be sure to discuss the following questions with the other members of your group:

- What might **cause** the appearance of the trees near rivers or streams to **change**?
- How can we **measure a change** over time?

Investigation Log

Our guiding question:

This is a picture of how we will set up the equipment:

We will collect the following data:

These are the steps we will follow to collect data:

I approve of this investigation proposal.

_____ _____
Teacher's signature   Date

Investigation 14. Movement of Water:
Why Can We See the Roots of Trees That Grow Near Rivers or Streams?

Collect Your Data

Keep a record of what you measure or observe during your investigation in the space below.

Analyze Your Data

You will need to analyze the data you collected before you can develop an answer to the guiding question. To do this, create a graph that shows the relationship between the cause and the effect.

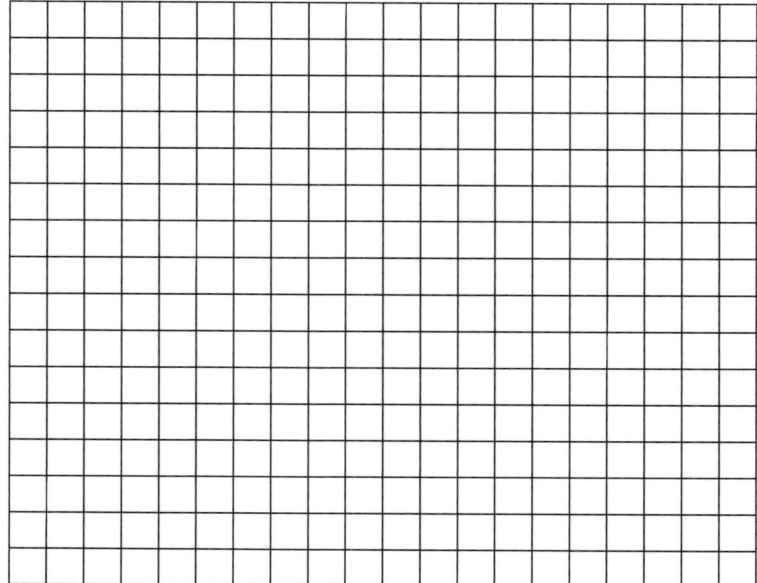

Student Workbook for Argument-Driven Inquiry in **Fourth-Grade Science**: Three-Dimensional Investigations

Investigation Log

Draft Argument

Develop an argument on a whiteboard. It should include the following:

- A *claim*: Your answer to the guiding question.
- *Evidence*: An analysis of the data and an explanation of what the analysis means.
- A *justification of the evidence*: Why your group thinks the evidence is important.

The Guiding Question:	
Our Claim:	
Our Evidence:	Our Justification of the Evidence:

Argumentation Session

Share your argument with your classmates. Be sure to ask them how to make your draft argument better. Keep track of their suggestions in the space below.

Ways to IMPROVE our argument …

Investigation 14. Movement of Water:
Why Can We See the Roots of Trees That Grow Near Rivers or Streams?

Draft Report

Prepare an *investigation report* to share what you have learned. Use the information in this handout and your group's final argument to write a *draft* of your investigation report.

Introduction

We have been studying _____ in class. Before we started

this investigation, we explored _____

_____

_____

We noticed _____

_____

_____

My goal for this investigation was to figure out _____

_____

_____

The guiding question was _____

_____

_____

Method

To gather the data I needed to answer this question, I _____

_____

_____

_____

Investigation Log

_____

_____

_____

I then analyzed the data I collected by _____

_____

_____

_____

Argument

My claim is _____

_____

_____

The graph below shows _____

_____

_____

_____

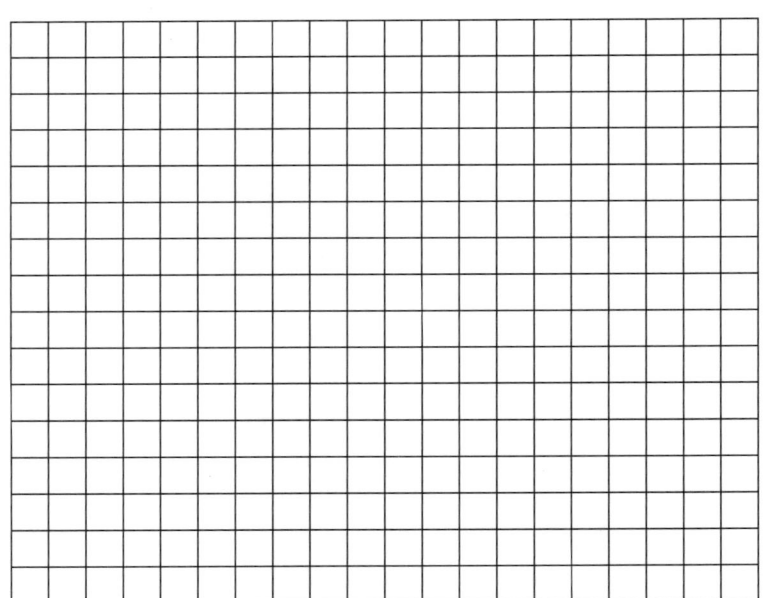

Investigation 14. Movement of Water:
Why Can We See the Roots of Trees That Grow Near Rivers or Streams?

This analysis of the data I collected suggests _____

_____

_____

_____

_____

This evidence is important because of several scientific concepts. The first one is _____

_____

_____

_____

_____

_____

_____

_____

_____

 **Review**

Your friends need your help! Review the draft of their investigation reports and give them ideas about how to improve. Use the *peer-review guide* that begins on the next page to guide your review.

Investigation Log

Peer-Review Guide

Section 1: The Investigation		Reviewer Rating	
1. Did the author do a good job of explaining what the investigation was about?	☐ No	☐ Almost	☐ Yes
2. Did the author do a good job of making the **guiding question** clear?	☐ No	☐ Almost	☐ Yes
3. Did the author do a good job of describing what he or she did to **collect data**?	☐ No	☐ Almost	☐ Yes
4. Did the author do a good job describing **how** he or she **analyzed** the data?	☐ No	☐ Almost	☐ Yes

Reviewers: If your group gave the author any "No" or "Almost" ratings, please give the author some advice about what to do to improve this part of his or her investigation report.

Section 2: The Argument		Reviewer Rating	
1. Does the author's claim provide a clear and detailed **answer** to the guiding question?	☐ No	☐ Almost	☐ Yes
2. Did the author support his or her claim with **scientific evidence**? Scientific evidence includes analyzed data and an explanation of the analysis.	☐ No	☐ Almost	☐ Yes
3. Does the **evidence** that the author uses in his or her argument **support the claim**?	☐ No	☐ Almost	☐ Yes
4. Did the author include enough **evidence** in his or her argument?	☐ No	☐ Almost	☐ Yes
5. Did the author do a good job of **explaining why the evidence** is important (why it matters)?	☐ No	☐ Almost	☐ Yes
6. Is the content of the argument **correct** based on the science concepts we talked about in class?	☐ No	☐ Almost	☐ Yes

Reviewers: If your group gave the author any "No" or "Almost" ratings, please give the author some advice about what to do to improve this part of his or her investigation report.

Continued

Investigation 14. Movement of Water:
Why Can We See the Roots of Trees That Grow Near Rivers or Streams?

Section 3: Mechanics	Reviewer Rating		
1. **Grammar:** Are the sentences complete? Is there proper subject-verb agreement in each sentence? Are there no run-on sentences?	☐ No	☐ Almost	☐ Yes
2. **Conventions:** Did the author use proper spelling, punctuation, and capitalization?	☐ No	☐ Almost	☐ Yes
3. **Word Choice:** Did the author use the right words in each sentence (for example, *there* vs. *their, to* vs. *too, then* vs. *than*)?	☐ No	☐ Almost	☐ Yes

Reviewers: If your group gave the author any "No" or "Almost" ratings, please give the author some advice about what to do to improve the writing mechanics of his or her investigation report.

General Reviewer Comments

We liked …

We wonder …

Investigation Log

Write Your Final Report

Once you have received feedback from your friends about your draft report, create your final investigation report in the space that follows.

Introduction

___

Method

___

Investigation 14. Movement of Water:
Why Can We See the Roots of Trees That Grow Near Rivers or Streams?

Argument

_____
_____
_____
_____
_____
_____
_____
_____
_____

_____
_____
_____
_____
_____

Investigation Log

Investigation Report Grading Rubric

Section 1: The Investigation	Missing	Somewhat	Yes
1. The author explained what the investigation was about.	0	1	2
2. The author made the **guiding question** clear.	0	1	2
3. The author **described** what he or she did to **collect data**.	0	1	2
4. The author described **how** he or she **analyzed** the data.	0	1	2
Section 2: The Argument	Missing	Somewhat	Yes
1. The claim includes a clear and detailed **answer** to the guiding question.	0	1	2
2. The author used **scientific evidence** to support the claim. Scientific evidence includes analyzed data and an explanation of the analysis.	0	1	2
3. The evidence **supports the claim.**	0	1	2
4. The author included enough **evidence** in his or her argument.	0	1	2
5. The author **explained why the evidence** is important.	0	1	2
6. The content of the argument is **correct**.	0	1	2
Section 3: Mechanics	Missing	Somewhat	Yes
1. **Grammar:** The sentences are complete. There is proper subject-verb agreement in each sentence. There are no run-on sentences.	0	1	2
2. **Conventions:** The author used proper spelling, punctuation, and capitalization.	0	1	2
3. **Word Choice:** The author used the right words in each sentence (e.g., *there* vs. *their, to* vs. *too, then* vs. *than*).	0	1	2

Teacher Comments

Here are some things I really liked about your report …	Here are some things I think you could do next time to make your report even better …

Total: _____ /26

Checkout Questions

Investigation 14. Movement of Water

The picture below shows a creek and the land around the creek. The squares in the creek are 30 cm on each side. Use this information to answer questions 1–3.

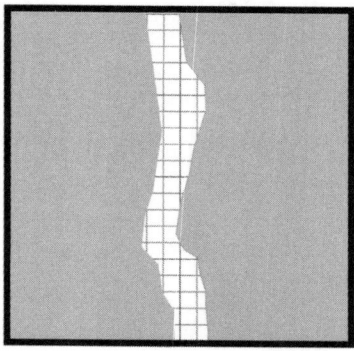

1. Which picture below (A or B) shows the least amount of soil erosion along the banks of this creek? Circle your choice.

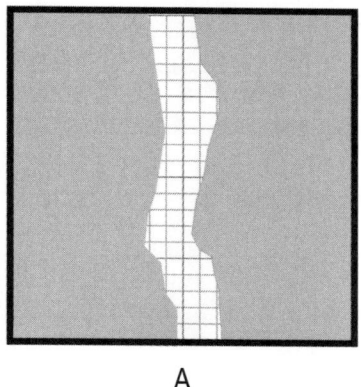
A

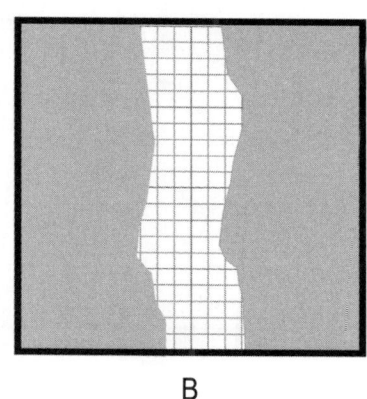
B

2. Which picture below (A or B) shows what the creek would look like after several days of heavy rain? Circle your choice.

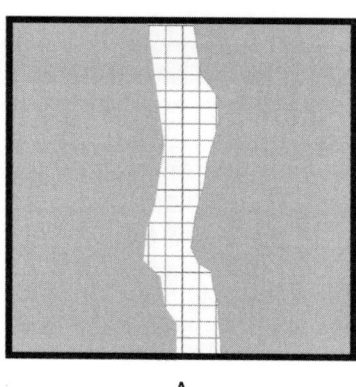
A

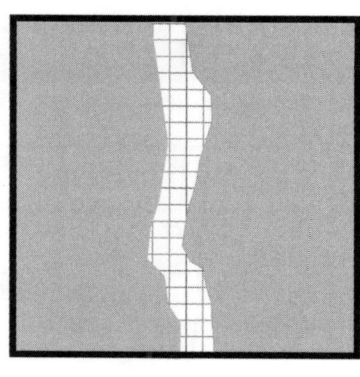
B

Checkout Questions

3. Explain your thinking. What *cause-and-effect relationship* from your investigation did you use to answer these questions?

___

Teacher Scoring Rubric for the Checkout Questions

Level	Description
3	The student can apply the core idea correctly in all cases and can fully explain the cause-and-effect relationship.
2	The student can apply the core idea correctly in all cases but cannot fully explain the cause-and-effect relationship.
1	The student cannot apply the core idea correctly in all cases but can fully explain the cause-and-effect relationship.
0	The student cannot apply the core idea correctly in all cases and cannot explain the cause-and-effect relationship.

Investigation Log

Investigation 15
Earth's Features: Why Do Large Waves Often Block the Entrance to Some Harbors in New Zealand?

Introduction

Many people use boats to go out on the ocean to work, to travel to different places, or for recreation. When people are not using a boat, they will often keep it (dock it) in a harbor, which is a sheltered body of water that is connected to an ocean. An example of a harbor is the one that is located at the mouth of the Whakatane River in New Zealand. Although Whakatane Harbour provides a place to dock a boat, it can be difficult for boaters to get in or out of it because there are often large waves that block the entrance to this harbor. (Note that in New Zealand some words are spelled differently than in the United States. For example, they use "Harbour" instead of "Harbor.") Your teacher will show a video of several boats attempting to enter this harbor despite the large waves. Keep track of what you observe as you watch the video and what you are wondering about in the boxes below.

Things I OBSERVED …	Things I WONDER about …

The large waves that often block the entrance to Whakatane Harbour are an example of a *natural hazard*. A natural hazard is any naturally occurring event that can cause harm to humans. Humans cannot prevent natural hazards, but we can take steps to warn people about them or to reduce how much

Investigation Log

damage they cause when they happen. Scientists therefore often study what causes natural hazards to happen because once they understand why and how these events happen, they can predict when and where the next one will occur. Engineers can also use an understanding about what causes a natural hazard to develop new tools or systems to help keep people safe. One way to learn more about the cause of a natural hazard is to study the features of the water and the land in areas where a natural hazard has happened or continues to happen.

Maps show where things are located. People can map the shapes and kinds of water and land in any area. One kind of map that people often use to describe the water and land at a specific location is called a *topographic map*. A topographic map describes the physical features of an area. These maps include natural features such as mountains, rivers, lakes, and valleys and manufactured features such as roads, dams, and cities. These maps also tell us the elevation of the land in an area. Another type of map that people often use to describe the water and land in a specific area is called a *nautical chart*. A nautical chart provides information about the depth of water and the shape of coastlines at a specific location (such as in or around a harbor). Both of these maps are useful for studying natural hazards because they make it possible to compare the features of the water and the land in different areas.

In this investigation, your goal is to figure out why the entrances to some harbors in New Zealand are often blocked by large waves and why the entrances to other harbors are not. You will need to use nautical charts to compare and contrast the features of the water and the land in and around these different harbors. Your teacher will show you how to use these nautical charts. You can use this method to figure out what causes the large waves that block the entrance of some harbors. Keep in mind that waves are caused by natural processes that are influenced by the features of the water and the land at a specific location.

Things we KNOW from what we read …	What we will NEED to figure out …

Investigation 15. Earth's Features:
Why Do Large Waves Often Block the Entrance to Some Harbors in New Zealand?

Your Task

Use what you know about patterns, structure and function, natural hazards, and maps to design and carry out an investigation to figure out why the entrances to some harbors in New Zealand are often blocked by large waves and why the entrances to other harbors are not.

The *guiding question* of this investigation is, **Why do large waves often block the entrance to some harbors in New Zealand?**

Materials

You will have access to nautical charts for the following New Zealand harbors during your investigation:

Harbors Often Blocked by Large Waves

- Kaipara Harbour
- Kawhia Harbour
- Manukau Harbour
- Raglan Harbour (Whaingaroa)

Harbors Rarely Blocked by Large Waves

- Anaura Bay Harbour
- Doubtless Bay Harbour
- Herekino Bay
- Mangonui Harbour
- Tokomaru Bay
- Tolaga Bay

Safety Rules

Follow all normal safety rules.

Plan Your Investigation

Prepare a plan for your investigation by filling out the chart that follows; this plan is called an *investigation proposal*. Before you start developing your plan, be sure to discuss the following questions with the other members of your group:

- How might the **structure** of the harbor be related to how it **functions**?
- What types of **patterns** might we look for to help answer the guiding question?

Investigation Log

Our guiding question:

We will collect the following data:

These are the steps we will follow to collect data:

I approve of this investigation proposal.

_____   _____
Teacher's signature   Date

Investigation 15. Earth's Features:
Why Do Large Waves Often Block the Entrance to Some Harbors in New Zealand?

Collect Your Data

Keep a record of what you measure or observe during your investigation in the space below.

Analyze Your Data

You will need to analyze the data you collected before you can develop an answer to the guiding question. To do this, create a table, a graph, or a picture in the space below.

Investigation Log

Draft Argument

Develop an argument on a whiteboard. It should include the following:

- A *claim:* Your answer to the guiding question.
- *Evidence:* An analysis of the data and an explanation of what the analysis means.
- A *justification of the evidence:* Why your group thinks the evidence is important.

The Guiding Question:	
Our Claim:	
Our Evidence:	Our Justification of the Evidence:

Argumentation Session

Share your argument with your classmates. Be sure to ask them how to make your draft argument better. Keep track of their suggestions in the space below.

Ways to IMPROVE our argument …

Investigation 15. Earth's Features: Why Do Large Waves Often Block the Entrance to Some Harbors in New Zealand?

Draft Report

Prepare an *investigation report* to share what you have learned. Use the information in this handout and your group's final argument to write a *draft* of your investigation report.

Introduction

We have been studying _____ in class. Before we started this investigation, we explored _____

_____

_____

We noticed _____

_____

_____

My goal for this investigation was to figure out _____

_____

_____

The guiding question was _____

_____

_____

Method

To gather the data I needed to answer this question, I _____

_____

_____

_____

_____

Student Workbook for Argument-Driven Inquiry in **Fourth-Grade Science:** Three-Dimensional Investigations

Investigation Log

_____
_____
_____

I then analyzed the data I collected by _____
_____
_____
_____

Argument

My claim is _____
_____
_____

The _____ below shows _____
_____
_____
_____

Investigation 15. Earth's Features:
Why Do Large Waves Often Block the Entrance to Some Harbors in New Zealand?

This analysis of the data I collected suggests _____

_____

_____

_____

_____

_____

_____

This evidence is important because of several scientific concepts. The first one is _____

_____

_____

_____

_____

_____

_____

_____

_____

_____

 **Review**

Your friends need your help! Review the draft of their investigation reports and give them ideas about how to improve. Use the *peer-review guide* that begins on the next page to guide your review.

Peer-Review Guide

Section 1: The Investigation		Reviewer Rating	
1. Did the author do a good job of explaining what the investigation was about?	☐ No	☐ Almost	☐ Yes
2. Did the author do a good job of making the **guiding question** clear?	☐ No	☐ Almost	☐ Yes
3. Did the author do a good job of describing what he or she did to **collect data**?	☐ No	☐ Almost	☐ Yes
4. Did the author do a good job describing **how** he or she **analyzed** the data?	☐ No	☐ Almost	☐ Yes
Reviewers: If your group gave the author any "No" or "Almost" ratings, please give the author some advice about what to do to improve this part of his or her investigation report.			

Section 2: The Argument		Reviewer Rating	
1. Does the author's claim provide a clear and detailed **answer** to the guiding question?	☐ No	☐ Almost	☐ Yes
2. Did the author support his or her claim with **scientific evidence?** Scientific evidence includes analyzed data and an explanation of the analysis.	☐ No	☐ Almost	☐ Yes
3. Does the **evidence** that the author uses in his or her argument **support the claim?**	☐ No	☐ Almost	☐ Yes
4. Did the author include enough **evidence** in his or her argument?	☐ No	☐ Almost	☐ Yes
5. Did the author do a good job of **explaining why the evidence** is important (why it matters)?	☐ No	☐ Almost	☐ Yes
6. Is the content of the argument **correct** based on the science concepts we talked about in class?	☐ No	☐ Almost	☐ Yes
Reviewers: If your group gave the author any "No" or "Almost" ratings, please give the author some advice about what to do to improve this part of his or her investigation report.			

Continued

Investigation 15. Earth's Features:
Why Do Large Waves Often Block the Entrance to Some Harbors in New Zealand?

Section 3: Mechanics	Reviewer Rating		
1. ***Grammar:*** Are the sentences complete? Is there proper subject-verb agreement in each sentence? Are there no run-on sentences?	☐ No	☐ Almost	☐ Yes
2. ***Conventions:*** Did the author use proper spelling, punctuation, and capitalization?	☐ No	☐ Almost	☐ Yes
3. ***Word Choice:*** Did the author use the right words in each sentence (for example, *there* vs. *their, to* vs. *too, then* vs. *than*)?	☐ No	☐ Almost	☐ Yes

Reviewers: If your group gave the author any "No" or "Almost" ratings, please give the author some advice about what to do to improve the writing mechanics of his or her investigation report.

General Reviewer Comments

We liked …

We wonder …

Investigation Log

Write Your Final Report

Once you have received feedback from your friends about your draft report, create your final investigation report in the space that follows.

Introduction

_____

_____

_____

_____

_____

_____

_____

Method

_____

_____

_____

_____

_____

_____

_____

_____

_____

_____

_____

Investigation 15. Earth's Features:
Why Do Large Waves Often Block the Entrance to Some Harbors in New Zealand?

Argument

Investigation Report Grading Rubric

Section 1: The Investigation	Missing	Somewhat	Yes
1. The author explained what the investigation was about.	0	1	2
2. The author made the **guiding question** clear.	0	1	2
3. The author **described** what he or she did to **collect data.**	0	1	2
4. The author described **how** he or she **analyzed** the data.	0	1	2

Section 2: The Argument	Missing	Somewhat	Yes
1. The claim includes a clear and detailed **answer** to the guiding question.	0	1	2
2. The author used **scientific evidence** to support the claim. Scientific evidence includes analyzed data and an explanation of the analysis.	0	1	2
3. The evidence **supports the claim.**	0	1	2
4. The author included enough **evidence** in his or her argument.	0	1	2
5. The author **explained why the evidence** is important.	0	1	2
6. The content of the argument is **correct.**	0	1	2

Section 3: Mechanics	Missing	Somewhat	Yes
1. **Grammar:** The sentences are complete. There is proper subject-verb agreement in each sentence. There are no run-on sentences.	0	1	2
2. **Conventions:** The author used proper spelling, punctuation, and capitalization.	0	1	2
3. **Word Choice:** The author used the right words in each sentence (e.g., *there* vs. *their*, *to* vs. *too*, *then* vs. *than*).	0	1	2

Teacher Comments

Here are some things I really liked about your report …	Here are some things I think you could do next time to make your report even better …

Total: _____ /26

256

National Science Teachers Association

Checkout Questions

Investigation 15. Earth's Features

The graph below shows how the depth of the water changes in two different harbors. Use this graph to answer questions 1 and 2.

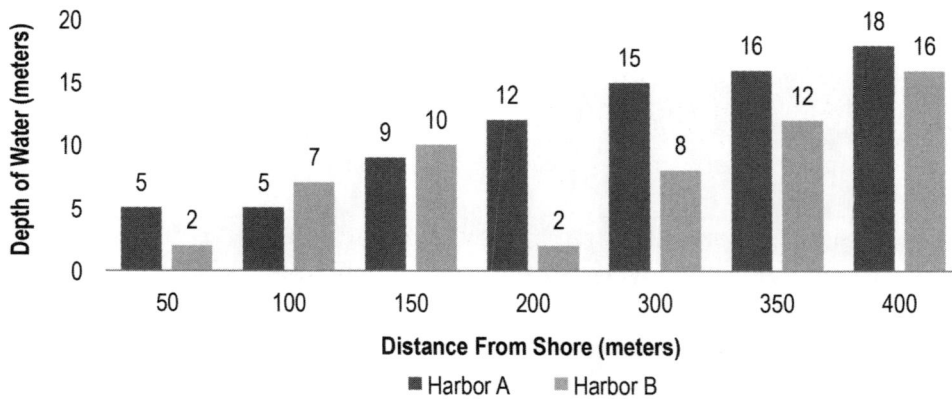

1. In which harbor would boaters need to watch out for large waves while they are entering or leaving?

    a. Harbor A

    b. Harbor B

    c. Both Harbor A and Harbor B

    d. Neither Harbor A nor Harbor B

2. Explain your thinking. What *pattern* from your investigation did you use to determine which harbor is likely to have large waves at its entrance?

_____
_____
_____
_____
_____
_____

Teacher Scoring Rubric for the Checkout Questions

Level	Description
3	The student can apply the core idea correctly in all cases and can fully explain the pattern.
2	The student can apply the core idea correctly in all cases but cannot fully explain the pattern.
1	The student cannot apply the core idea correctly in all cases but can fully explain the pattern.
0	The student cannot apply the core idea correctly in all cases and cannot explain the pattern.

IMAGE CREDITS

All images in this book are stock photos or courtesy of the authors unless otherwise noted below.

Investigation 12

Checkout Question illustration below question 1: Klaus Rassinger und Gerhard Cammerer, Museum Wiesbaden, Wikimedia Commons, CC BY-SA 3.0, *https://commons.wikimedia.org/wiki/File:Lama_guanicoe_02_MWNH_820.jpg*

Checkout Question illustration below question 2: Klaus Rassinger und Gerhard Cammerer, Museum Wiesbaden, Wikimedia Commons, CC BY-SA 3.0, *https://commons.wikimedia.org/wiki/File:Trachypithecus_auratus_02_MWNH_115.jpg*

Checkout Question illustration below question 3: Klaus Rassinger und Gerhard Cammerer, Museum Wiesbaden, Wikimedia Commons, CC BY-SA 3.0, *https://commons.wikimedia.org/wiki/File:Ursus_maritimus_02_MWNH_420.JPG*